Lessons from Forest Decentralization

The Earthscan Forest Library

Series Editor: Jeffrey A. Sayer

The Forest Certification Handbook 2nd edition
Ruth Nussbaum and Markku Simula

The Forest Landscape Restoration Handbook
Jennifer Rietbergen-McCracken, Stewart Maginnis and Alastair Sarre (eds)

Forest Quality: Assessing Forests at a Landscape Scale
Nigel Dudley, Rodolphe Schlaepfer, Jean-Paul Jeanrenaud, William Jackson and Sue Stolton

Forests in Landscapes: Ecosystem Approaches to Sustainability
Jeffrey A. Sayer and Stewart Maginnis (eds)

Forests, People and Power: The Political Ecology of Reform in South Asia
Oliver Springate-Baginski and Piers Blaikie (eds)

Illegal Logging: Law Enforcement, Livelihood and the Timber Trade
Luca Tacconi (ed)

Lessons from Forest Decentralization: Money, Justice and the Quest for Good Governance in Asia-Pacific
Carol J. Pierce Colfer, Ganga Ram Dahal and Doris Capistrano (eds)

Logjam: Deforestation and the Crisis of Global Governance
David Humphreys

Plantations, Privatization, Poverty and Power: Changing Ownership and Management of State Forests
Mike Garforth and James Mayers (eds)

Policy that Works for Forests and People: Real Prospects for Governance and Livelihoods
James Mayers and Stephen Bass

The Politics of Decentralization: Forests, Power and People
Carol J. Pierce Colfer and Doris Capistrano

The Sustainable Forestry Handbook 2nd edition
Sophie Higman, James Mayers, Stephen Bass, Neil Judd and Ruth Nussbaum

Lessons from Forest Decentralization

Money, Justice and the Quest for Good Governance in Asia-Pacific

Edited by
Carol J. Pierce Colfer,
Ganga Ram Dahal
and
Doris Capistrano

earthscan
from Routledge

First published by Earthscan in the UK and USA in 2008

For a full list of Earthscan publications please contact:
Earthscan
2 Park Square, Milton Park, Abingdon, Oxfordshire OX14 4RN
711 Third Avenue, New York, NY, 10017, USA

First issued in paperback 2014

Earthscan is an imprint of the Taylor & Francis Group, an informa business

A catalogue record for this book is available from the British Library

Library of Congress Cataloging-in-Publication Data
Lessons from forest decentralization : money, justice, and the quest for good governance in Asia-Pacific / edited by Carol J. Pierce Colfer, Ganga Ram Dahal, and Doris Capistrano.
p. cm.
ISBN 978-1-84407-555-3 (hardback)
1. Forest management—Asia. 2. Forest management—Asia—Citizen participation. 3. Forest policy—Asia. I. Colfer, Carol J. Pierce. II. Dahal, Ganga Ram. III. Capistrano, Doris.
SD219.L47 2008
333.75095—dc22

2007041629

ISBN-13: 978-1-84407-555-3 (hbk)
ISBN-13: 978-0-415-84561-8 (pbk)

Typeset by JS Typesetting Ltd, Porthcawl, Mid Glamorgan
Cover design by Susanne Harris

Contents

List of Figures, Tables and Boxes

Figures

Tables

Boxes

List of Contributors

Seema Arora-Jonsson PhD is an Indian rural development specialist and Research Fellow at the Centre for Gender Research, Uppsala University and Department of Urban and Rural Studies, Swedish University of Agricultural Sciences, Sweden.

Domingo Bacalla is Filipino Chief of the Community-Based Forest Management Division of the Forest Management Bureau, within the Department of Environment and Natural Resources, in Manila, Philippines.

Yati Bun is from Papua New Guinea and is Executive Director of the Foundation for People and Community Development Inc, in Boroko, Papua New Guinea.

Byoung Il Yoo PhD is a Korean forester seconded to the Center for International Forestry Research in Bogor, Indonesia.

Doris Capistrano PhD is a Filipina economist and Director of the Programme on Forest Governance at the Center for International Forestry Research in Bogor, Indonesia

Apsara Chapagain is from Nepal and is an active member of FECOFUN, the Federation of Community Forestry Users, Nepal.

Carol J. Pierce Colfer PhD is an American anthropologist and Principal Scientist at the Center for International Forestry Research in Bogor, Indonesia.

Ganga Ram Dahal PhD is a Nepali researcher and Postdoctoral Fellow at the Center for International Forestry Research in Bogor, Indonesia.

Efransjah PhD is a Malaysian Chief Technical Advisor to UNDP/GEF, also working with Forest Research Institute Malaysia and Ministry of Natural Resources and Environment, Malaysia.

Ferdinand Esguerra is from the Philippines and works as the EcoGov Regional Coordinator for Central and Southern Mindanao, in the Philippines.

Robert Fisher PhD is an Australian currently working as a Senior Researcher for the Australian Mekong Resource Centre at the University of Sydney in Australia.

Yayoi Fujita is a Japanese researcher working in the Lao People's Democratic Republic.

Ernesto Guiang PhD is a Filipino and Chief of Party on a project sponsored by Development Alternatives, Inc, on Philippines Environmental Governance, in Pasig City, Philippines.

Petrus Gunarso PhD is an Indonesian currently working as a Coordinator for the Malinau Research Forest, East Kalimantan, Indonesia.

Che Hashim bin Hassan MSc is a Malaysian forester, and Director of Pahang State Forestry Department, Malaysia.

Kadjatmiko is an Indonesian Director of Balancing Fund, Directorate General of Budget and Fiscal Balance, Ministry of Finance Indonesia.

Kaisone Phengsopha is a Laotian lecturer at the Faculty of Forestry, National University of Laos and involved in mobile environmental education programme in Laos.

Christian Küchli is a Swiss forester working for the Swiss government at BAFU, Bundesamt fur Umwelt (Federal Department of Environment).

Liu Dachang PhD is a Chinese conservation and Forest Scientist with The Nature Conservancy, China Programme, based in Kunming, People's Republic of China.

Moira Moeliono PhD is an Indonesian trained in forestry and social geography working at the Center for International Forestry Research in Bogor, Indonesia.

Tan Quang Nguyen PhD is a Vietnamese working as an Independent Researcher in the field of forestry.

Colin Lockhart O'Loughlin has a PhD in forest hydrology from the University of British Columbia and works as a consultant in forestry and science based in Christchurch, New Zealand.

Lesley Potter PhD is an Australian geographer and Visiting Fellow in the Department of Human Geography, Research School of Pacific and Asian Studies in the Australian National University, Australia.

Kresno D. Santosa is an Indonesian working as a Research Assistant in Malinau Research Forest, East Kalimantan, Indonesia.

Yunety Tarigan is an Indonesian, working as a Research Assistant at the Centre for International Forestry Research in Bogor, Indonesia.

Foreword

We heartily welcome the preparation of this book on forest governance and decentralization in Asia and the Pacific. The chapters reflect the shared development of thinking among countries in Asia and the Pacific and synthesis of the main points emerging from the Workshop on Forest Governance and Decentralization, conducted 4–6 September 2006 in Yogyakarta, Indonesia. This important event helped broaden understanding of decentralization among many parties throughout the region.

From our experience and sharing information on the issues of decentralization, we finally recognize that decentralization is a dynamic process that includes constant learning and experimentation. It needs to be phased in gradually. It requires building consensus through an open, transparent and inclusive process; participatory decision-making; institutional, technical and human capacity-building; provision of adequate financial resources and incentives for investment; tailoring of objectives to local contexts; and developing the flexibility to adapt to different situations and changing circumstances. Priority in decentralization must then be accorded to empowerment and capacity-building of the local communities so that they can effectively manage their natural resources. Decentralization should not simply transfer the power, authority and burden of management – it also means sharing responsibilities and creating net positive benefits for local communities.

We would like to extend our thanks to the organizing committee, whose members facilitated and organized the workshop so well. Our sincere thanks also go to the donors who have contributed to the event – namely the International Tropical Timber Organization, the Multi-stakeholders Forestry Programme of the UK Department for International Development, Inter-cooperation of Switzerland, the Strengthening Management Capacity Programme of Deutsche Gesellschaft für Technische Zusammenarbeit (GTZ), and the UN Food and Agriculture Organization. We hope that this cooperation will continue in the future to achieve sustainable forest management. We are also grateful to all participants in the Workshop on Forest Governance and Decentralization who were actively involved and have contributed to the discussion. Our most sincere thanks and gratitude also go to the Provincial Government of Yogyakarta; Heads of the Districts of Gunung Kidul, Purworejo and Wonogiri; Perum Perhutani; the Dean of the Faculty of

Forestry of Gajah Mada University; and the Forest Students Associations of Gajah Mada University and Bogor Agriculture Institute. The people of Yogyakarta are also gratefully acknowledged for hosting and making excellent arrangements for this workshop.

We have found this workshop inspiring not only because of its recommendations but also for the lessons brought from each country from its path across the dynamic terrain of decentralization.

M. S. Kaban
Minister of Forestry
Jakarta, October 2006

Acknowledgements

Although each author has additional acknowledgements, we wanted to mention some to whom we *all* are grateful. Our biggest debt in the development of this book has been to the Asia Forest Partnership, whose members initially conceived the idea to bring experts on Asia and Pacific together to discuss forest governance. The Indonesian Government, which hosted the Yogyakarta Workshop on Forest Governance and Decentralisation in Asia and the Pacific, also supported the effort financially, as did a number of partner organizations (the International Tropical Timber Organization, Deutsche Gesellschaft fur Technische Zusammenarbeit, the United Kingdom's Department for International Development and Swiss InterCooperation), along with Japan and RECOFTC, both of whom supported participants. We also thank the Philippine Government, the Food and Agriculture Organization of the United Nations (FAO), and Switzerland's Federal Office for the Environment (BAFU), all of which were instrumental in moving the idea for this workshop and book forward. The chapters in this book were all reviewed by an all-volunteer Review Panel, to which we express our sincere gratitude for enhancing the quality and accuracy of our effort. We also thank our own institution, the Center for International Forestry Research, which has provided excellent support – administrative, logistical, and intellectual.

Although it is impossible to identify all those who have contributed, we do want to single out a few who have put forth tremendous effort and been instrumental in the success of the endeavour. Specifically, we could not have done without the able and consistent work of Tonny Soehartono, Bun Purnama and Thomas Nifinluri (all of the Indonesian Ministry of Forestry), Patrick Durst of FAO, Emmanuel Zemeka and Hiras Subaditar of ITTO, Jurgen Blaser and Annette Luthi of InterCooperation, and Mike Nurse of Bangkok's Regional Community Forestry Training Center (RECOFTC), all of whom helped conceptually and practically in developing the topics to be addressed, the authors who would address them, the participants in the Yogya Workshop and the format in which our discussions would take place.

At CIFOR, we have all relied on the able administrative coordination of Yunety Tarigan, and as always we are grateful to Gideon Suharyanto and Rahayu Koesnadi for their loyal help, whenever needed. Yayan Indriatmoko, a 'native son', provided valuable input in field site selection and photographic expertise during the Yogyakarta Workshop in September 2006. Nathaniel Gerhart, a young doctoral student who worked with CIFOR informally and was tragically killed in a motorcycle accident in 2007, ably led sessions and contributed useful insights. Ambar Liano, Dina Hubudin, Heru Komarudin, Ahmad Dermawan and Nugroho Adi Utomo supplemented all their efforts at the workshop. We thank Sally Atwater for her meticulous care in correcting the English of non-native speakers and for making more elegant prose for the rest of us. We are further indebted to the people at Earthscan (Rob West, Alison Kuznets, Gudrun Freese and Hamish Ironside), who have coordinated all the post-editing steps with patience, equanimity and helpfulness.

Ultimately, of course, at some level, our deepest gratitude must go to those people in government, business, and communities all over Asia and the Pacific that we have troubled with our questions and observations, and who have helped us gain insights into the processes that characterize forest governance. We hope that their time was well spent.

1

Setting the Stage: Money and Justice in Asia and Pacific Forests

Carol J. Pierce Colfer, Ganga Ram Dahal and Moira Moeliono

When the world's population was small and forests covered a larger portion of the globe, forest governance was probably not a very significant problem. But with increasing population and increasing consumption of the Earth's resources over the past century, forest governance has become a burning issue. Disputes abound, passions flare, fortunes are won and lost, and wars can even result from issues of forest governance. Partially in response to these problems, a popular move in recent decades has been to decentralize or devolve forest governance from central governments to lower levels of government, to civil society or to private hands (see, for example, Contreras, 2003b; Colfer and Capistrano, 2005; Ribot and Larson, 2005).

One common rationale for such changes – particularly within governments – is the idea that decentralized governance will be cheaper. A more idealistic argument – common in more academic writings and in development circles – is that decentralized/devolved modes of governance will be closer to the people and thus more amenable to the checks and balances that can control the excesses of government officials and also make them more responsive to the needs of the people. This in turn would enhance the effectiveness and efficiency of service delivery and help maintain equity and social justice (Burns et al, 1994; Turner and Hulme, 1997). This view, in the long run, is probably correct. However, there have been many bumps along the road to effective devolution in most countries, and Asia-Pacific countries – with which we are concerned here – are no exception. Many well-researched problems centre around, among other things, the division of labour and benefits between levels of government; the willingness of higher levels to grant authority to lower levels in practice; the complexity of forest-related requirements that communities are unable to fulfil (such as forest management plans); and the lack of institutional capacity and financial resources at the local level to carry out the devolved responsibilities.

This book deals with these issues and takes the next step by looking more deeply into some topics that have not been particularly well addressed in the decentralization and devolution literature, but which seriously affect the processes of devolution in the Asia-Pacific region. The Asia Forest Partnership, which initially proposed the analyses provided here, had particular interest in illegal logging, forest fires and rehabilitation. The Indonesian government, with whom we have collaborated throughout, was particularly interested in the financial implications of decentralization. From a theoretical point of view, Colfer, Dahal and Capistrano were interested in the relationships between devolution processes and equity, corruption, tenure, and land-use planning. All the parties involved were interested in looking forward, learning from and building on previous successes and failures in the region, and developing plans for collaboration and improved forest governance in the future. But before turning to the abundant and timely case material in this book, we first want briefly to introduce the major work already available on this topic (further supplemented in Chapter 3).

Recent Regional Research on Forest Governance

There has been considerable high-level interest from governments and donors in decentralized forest governance over the past decade, and many of these efforts have resulted in excellent published analyses. Beginning in 1997, Wollenberg and Edmunds developed and implemented a project called 'Creating Space for Local Forest Management'. Their purpose was to examine the impacts of the devolution process at the local level, with a whole range of partners in India, China and the Philippines (see, for example, Wollenberg et al, 2001; Contreras, 2003a, b; Edmunds and Wollenberg, 2003). Not long after this effort began, the Food and Agriculture Organization of the UN (FAO) and the Regional Community Forestry Training Center (RECOFTC) held the 'International Seminar on Decentralization and Devolution of Forest Management in Asia and the Pacific' in Davao, Philippines (1998; reported in Enters et al, 2000, and in Chapter 3, this volume). This meeting brought together representatives from many countries in Asia to share their experiences with this process.

In 1999, the Center for International Forestry Research (CIFOR) established the 'Local People, Devolution and Adaptive Collaborative Management' (ACM) programme, which addressed devolution, among other things, in 11 countries, 4 of which were in Asia and the Pacific (Indonesia, Kyrgyzstan, Nepal and the Philippines). This programme attempted to strengthen efforts to involve local communities in better forest management and governance, while observing and analysing the process. The study communities (and also later, to some degree, districts) made good progress in improving equity; strengthening community and district communication, negotiation, and conflict management skills; improving local forest-related regulations; and strengthening both community knowledge of the regulations and commitment to following and enforcing them (Hartanto et al, 2003; Anau et al, 2005; Dangol, 2005; Diaw and Kusumanto, 2005).

Meanwhile, within Indonesia, there has been an enormous amount of attention paid to the decentralization process. The 2002 collection by Colfer and Resosudarmo describes the situation shortly after Soeharto's fall, when decentralization debates were in the air (see especially Bennett, 2002; Kartodihardjo, 2002; Lindayati, 2002; Resosudarmo and Dermawan, 2002; Wrangham, 2002; and the collection by Lynch and Harwell, 2002, which summarizes studies by many mainly NGO partners). CIFOR's Ida Aju Pradnja Resosudarmo then initiated a study of decentralization in seven Indonesian provinces in the early 2000s, each resulting in a case study (Barr et al, 2001; Casson, 2001a, b; McCarthy, 2001a, b; Potter and Badcock, 2001; Obidzinski and Barr, 2003, synthesized in Barr et al, 2006). In 2003, CIFOR continued work on decentralization in Indonesia, moving towards action research at the district level in five Indonesian provinces (Soetarto et al, 2003; McGrath et al, 2005; Ngakan et al, 2005; Samsu et al, 2005; Sudirman et al, 2005; Tokede et al, 2005; Yasmi et al, 2005). Komarudin, who participated in this work, then coordinated the conduct of a project that used participatory action research to look at decentralization issues at both community and district levels in two Jambi districts of Sumatra (Komarudin et al, 2007).

The changing face of forest governance globally led us in 2004 to cooperate with the host governments of Indonesia and Switzerland and a large number of other partners in organizing the Interlaken Workshop in Switzerland on the same subject. This global meeting brought together people from government, NGOs, academia, projects and communities to discuss the ways that decentralization and devolution had played out in their countries (described in Colfer and Capistrano, 2005; see also Gregersen et al, 2004). The success of this workshop prompted the same sponsors to conduct a smaller meeting in Indonesia the following year (described in Departemen Kehutanan, 2005), and to organize and sponsor a follow-up regional workshop, in 2006, focusing on Asia and the Pacific – the findings reported here. These various studies and observations led to our commitment to continue looking at the issues of equity, corruption and land tenure, as part and parcel of more authentically devolving forest governance. In the remainder of this chapter, we highlight some important issues that permeate both the chapters in this collection and the literature on the process of devolution of forest management in Asia and the Pacific (see the Annex on page 15 for a brief description of the Yogyakarta Workshop itself and the associated field trips). We conclude by explaining the structure of the book.

Central Issues in Forest Governance[1]

We have already mentioned the various interests that prompted this book – illegal logging, forest fire and rehabilitation were of central interest to the Asia Forest

Partnership; the Indonesian hosts emphasized roles, responsibilities and rights, institutional analyses, and the financial implications of decentralization; and equity, corruption, land tenure and land-use planning were of central concern from the research perspective. In early interaction with the authors, we mentioned our interest in all these themes. Wanting, however, to have their genuine views on the important issues in their respective countries, we did not hold the authors to a strict set of topics. We emphasized our interest in hearing their views and their analyses of what was occurring in the countries they were writing about – thus the wide-ranging topics covered in this collection.

Here, we introduce a few key issues that are treated more fully in this book, looking briefly at equity and gender; corruption and elite capture; tenure; finance; and bureaucratic complexity, balance and effectiveness. These issues and more are discussed further in the final chapter.

Equity and gender

Although there have been a number of analyses documenting the lack of involvement of women in formal decentralization efforts in many countries (for example Dangol, 2005, on Nepal; Wiliam-de Vries and Sutarti, 2006, on Indonesia; Sarin et al, 2003, on India), the topic remains under-addressed in conventional decentralization literature and in formal governmental efforts to involve women. Nevertheless, involving women more meaningfully, rare as it has been, has led to better forest management in some cases. A recent and exhaustive analysis by Agrawal et al (2006) examined the impacts of women's participation in forest management in the Indian state of Madhya Pradesh. They looked statistically at 641 village-level forest protection committees, comparing those with more and less involvement of women, and found that outcomes such as forest regeneration and control of illegal grazing and felling all benefited significantly from women's involvement. Women's participation was, for instance, associated with a 24 per cent increase in the probability of controlling illicit grazing and a 28 per cent increase in control of illicit timber felling (p33). Sarin et al (2003) also provide very thorough qualitative analyses in three Indian contexts in which the active roles of women in forest management are associated with positive effects on forest conditions and livelihoods.

Our goal has not been only to provide more evidence that women are involved in forests and therefore somehow deserve to be included in forest management. Instead, we wanted to move forward discussion that would lead to real changes in how decentralization is done – so that women (and other marginalized groups) are more meaningfully integrated into the processes that are being devised to make decentralization a useful reality (see also the recent collection by Vernooy, 2006).

CIFOR, and the NGOs Gita Buana and PSHK-ODA, have been working closely with villagers and district officials in two Indonesian districts in Jambi,

Sumatra, to develop mechanisms that will more effectively involve local women in forest-related policymaking. In Bungo District, local women have worked with men to legalize their own community management of part of their traditional area (*hutan adat*, or customary forest) that is formally part of Indonesia's national forest estate (Marzoni, 2006; Munggoro et al, 2006; Permatasari, 2007). In Bungo and Tanjung Jabung Barat Districts, another group has been working with local women to strengthen collective action, in the hope of increasing women's involvement in forest-related income-generating projects (Siagian et al, 2006). Still others have been working at the district level to raise policymakers' awareness of this issue (Wiliam-de Vries, 2006; Wiliam-de Vries and Sutarti, 2006). In another part of Indonesia, Danau Sentarum in West Kalimantan, women are gaining confidence to get involved in discussions on the future of the national park (Yuliani et al, 2006). In line with these interests, this collection includes considerable attention to this topic (see in particular Chapter 4).

Corruption and elite capture

Corruption is another topic that has been somewhat taboo in formal decentralization contexts (as well as other contexts!). Yet it pops up again and again in informal conversation. In Indonesia, it is common to hear that Soeharto's rampant corruption has simply been 'devolved' to the 434 district heads throughout the country and continues unabated (cf. Véron et al, 2006, for India). While that is probably not entirely fair, the fact remains that corruption still plagues the populace.[2] And Indonesia is no exception; in fact, corruption can be so ingrained that the ordinary citizen does not recognize it. Sarin et al (2003) and Baviskar (2005) on India, and Contreras (2003a) and Dahal (2006) on the Philippines, provide numerous examples where government officials and local elites engage in corrupt practices that adversely affect both local forests and local livelihoods. Liu and Edmunds (2003), writing about China, note that, in efforts to reduce corruption, much depends on the degree to which local government officials can be held accountable for their actions by local farmers. Capistrano and Colfer (2005) identify this as a global problem, arguing that the most effective way to combat it is strong civil society (also reiterated by Larson and Ribot, 2005).

Illegal logging is often closely linked with corruption, and operates at various levels. At one level powerful elites drive the activity; at another local people may seize any opportunity to make a living; and in the middle is the government apparatus which either contributes to or constrains the practice.

Decentralization in Indonesia has also enabled the emergence of 'unofficial' levies on forest products at district and even sub-district levels, which makes illegal logs slightly more legal and contributes to the revenue of local governments (Barr et al, 2006). Recently there have been anti-money-laundering laws put in place, which hopefully will contribute to addressing these issues (for example Law 15/2002, partially revised by Law 25/2003; forestry crime is mentioned in Law 25/2003). CIFOR teams have also had some success at the village level in controlling elite capture through local collective action (Komarudin et al, 2007).

Tenure

Tenure issues, discussed at greater length in Chapter 3, continue to plague decentralization processes, as governments try (or fail to try) to sort out conflicting claims to forest lands. There is a large variation in forest ownership across the region, with government landownership dominant. Where government control has been strong and forests valuable, handing over control is especially hard. This has been a particular problem in Indonesia (Siswanto and Wardojo, 2005), where forest land remains legally under the control of the state. On the other hand, local communities have been planting trees for decades. The bureau of statistics estimates that community woodlots cover approximately 1,560,229 hectares with a standing stock of 39,564,003 cubic metres of mixed species. Most of what has been captured in these statistics is concentrated in Java and includes bamboo, teak, mahogany, pines, *Paraserianthes* and rosewood with a production potential of about 19 million cubic metres (Anonymous, 2005). To 'protect' these lands from state interference, however, local people refer to them as woodlots or tree gardens, not forest. On the outer islands, reform and decentralization encouraged local communities to reclaim large tracts of forest land as local governments took control over forest administration. This de facto decentralization has led to increased deforestation and conflict. While districts are slowly learning to adjust and in some cases some form of forest governance has emerged, forest regulations in 2003, and the revised decentralization law of 2004, recentralized many aspects of forest administration, so that forest land remains under central government control. Indeed, these changes were presaged in a 2002 Governmental Regulation (*Peraturan Pemerintah* No 34 of that year), and even earlier in the 1999 Forestry Law, which maintained a strong central flavour, contradicting the 1999 decentralization law (Resosudarmo, 2005). As a result, tenure for forest-dwelling communities remains insecure.

Financial implications

This issue surfaced as a bone of contention in Indonesia as soon as decentralization became law (Resosudarmo and Dermawan, 2002). These authors already described a 'tug of war' between the regions and the centre during the first year of decentralization in Indonesia (2001), part of which was financial. And more recently, Siswanto and Wardojo (2005) and Resosudarmo et al (2006) have described its ongoing relevance. Makapukaw and Mirasol (2005), writing from a bottom–up perspective, identified the difficulties of getting sufficient financial resources at the lower levels of the Philippine system as a major problem.

Another related and recurring topic pertains to devolution of responsibilities without the means of implementation. Such means can be related to finances or to authority. Mearns (2005) gives a good example of this situation in Mongolia, where herders have a strong traditional ethic of open access (as is the case in many forests of the world). Mongolia has decentralized politically, but not administratively or financially, and local people express dissatisfaction with the degree to which their needs and aspirations are addressed by the government. Malysheva (2005) outlines the lack of financial resources at local levels in Russia and the adverse effects this has on environmental protection.

In many countries there is a lack of clarity about responsibilities and rights or authority – with serious financial implications. Makapukaw and Mirasol (2005), for instance, in describing efforts to improve the decentralized management of the protected area of Mt Kitanglad Natural Park, complain that government needs to harmonize both conflicting laws and the roles of different management institutions.

Bureaucratic balance, complexity and effectiveness

There continues to be active debate about the efficacy of decentralization processes, with some authors clearly nostalgic for a return to stronger central control (for example Malysheva, 2005; Yoo, this volume). Many authors (for example, Liu and Edmunds, 2003; Capistrano and Colfer, 2005; Larson and Ribot, 2005; Véron et al, 2006) note the important roles that central government typically continues to play when decentralization is effective.

Continuing problems exist at the local level in a number of countries with complex bureaucratic requirements (for example overly technical management plans). Many decentralization processes involve efforts to devolve some authority to communities. Yet these efforts are often plagued by bureaucratic steps that are so complex that even sophisticated urbanites have trouble understanding them (cf. Baviskar, 2005).

In Indonesia, for instance, *hutan kemasyarakatan* (social forestry) involves a difficult and long procedure with uncertain outcomes, dependent on the results of a probation period of 2–5 years that is to be extended to 25 years only after several more administrative requirements are fulfilled, including the transformation of the farmers' group into a formal cooperative (ministerial decree SK 31, 2001; see also Bun, 2006, on similar problems in Papua New Guinea).

An important dimension along which Asian and Pacific countries differ is the degree to which governments are able to enforce their laws. China and Korea have strong traditions of central control, going back millennia; New Zealand, on the other hand, has a strong, democratic system that is able, by and large, to implement its laws.

In contrast, India, Indonesia, Lao PDR, Malaysia, Papua New Guinea and the Philippines have had persistent difficulties enforcing their own regulations.

Lack of funds and difficult terrain have been constraints in many areas. Recency of independence from colonial rulers, as well as inappropriateness and lack of adaptation of colonial rules are likely to be additional factors.

STRUCTURE OF THE BOOK

We begin in this book with a historical discussion by Christian Küchli, who in Chapter 2 introduces Switzerland's commitment to the idea that effective forest sector governance is a key to sustainable forest management in Asia and the Pacific. Chapter 3, by Lesley Potter, introduces the most significant, tenure-related processes recently underway in Asia and the Pacific (also emphasizing gender issues). She includes boxes on important trends in Papua New Guinea and New Zealand (with Yati Bun and Colin O'Loughlin respectively).

The remaining chapters proceed from South Asia (Chapters 4 and 5), through East Asia (Chapters 6 and 7), to Southeast Asia (Chapters 8–12), which dominates the book. The topics identified as central in the workshop – roles, responsibilities and rights; institutional analyses, including organizations, capacity-building and legal aspects; and financial implications of decentralization – are addressed in these analyses.

In Chapter 4, Seema Arora-Jonsson looks at forest management by starting with the collective action of women who live in forests in India and Sweden. Her analysis emphasizes the differences in ways men and women organize themselves and the differences in their forest-related goals. She stresses the significance of attending to gender issues in trying to plan and implement more equitable decentralization of forests.

In Chapter 5, Ganga Ram Dahal and Apsara Chapagain take us to Nepal, looking at how local-level mobilization of community user groups into an effective national network can affect the implementation of decentralization. The chapter begins with an introduction to Nepal's forest decentralization process – perhaps the best known in the world. It then describes the unique development of FECOFUN (Federation of Community Forest Users of Nepal), emphasizing both the network's considerable accomplishments and some of the remaining problems that the country and FECOFUN continue to face.

Turning to East Asia, Liu Dachang, in Chapter 6, provides a fascinating diachronic account of the various institutional elements of decentralization in China, addressing tenure, the rights to dispose of forest property, and the distribution of benefits and income from the forest. Chapter 7, by Byoung Il Yoo, sets the historical context for what is termed 'autonomy' in Korea, showing how the various responsibilities have been divided between the central and local levels of government and emphasizing forest fires and the implications of autonomy for their control.

Moving to Southeast Asia, in Chapter 8, Yayoi Fujita and Kaisone Phengsopha share their experience in the Lao People's Democratic Republic, which is considered by many to have some of the most progressive policies in Southeast Asia. These authors emphasize the differences in implementation between donor-funded projects and normal governmental implementation, indicating areas for improvement. Chapter 9, by Efransjah, Gunarso, Hassan and Santosa, compares consultative planning processes, deemed essential for effective forest governance, in Malaysia and Indonesia.

Chapter 10, by Kadjatmiko, describes the extreme centralization that characterized the Soeharto regime in Indonesia and the devolution of authority intended to meet regional aspirations, improve the fiscal structure, and mobilize regional and national revenues. Chapter 11, by Ernesto Guiang, F. Esguerra and Domingo Bacalla, stresses the importance of finance, providing a historical description of Philippine forestry policy and the growing disconnect between the intentions with which these policies were developed and the realities and impacts on the ground. Tan Quang Nguyen, from Vietnam, describes in Chapter 12 the radical reform of the forestry sector that has occurred over the last 25 years, analysing Vietnam's bold efforts to devolve forest management to local people from a micro-financial perspective.

The final chapter (Chapter 13), by Capistrano, synthesizes what we have learned about forest governance and decentralization in Asia and the Pacific.

As will become clear in the chapters that follow, the diversity that characterizes the countries of Asia and the Pacific reinforces a conclusion we and others have come to before: the search for 'best practices' in the devolution of forest governance requires careful attention to history, culture and local contexts. But these chapters, despite their diversity, do provide illustrative examples of some approaches that have worked and others that have created or exacerbated problems. Policymakers and scholars can mine these pages for useful ideas and warnings for their efforts to improve governance, both within the region and elsewhere.

Notes

1 In this book, we focus on Asia and the Pacific. There is a large body of literature on decentralization and devolution from Africa (for example Oyono, forthcoming; Ribot and Larson, 2005; Sithole, 2005) and South America (Larson, 2005; Pacheco, 2002; Kaimowitz et al, 2000, among many others) which we cannot cover here, though we do refer to this rich literature from time to time.

2 Indonesia's status has improved slightly on Transparency International's corruption scale. It is now 2.0, or 133rd out of 146 countries, www.transparency.org/publications/gcr/download_gcr/download_gcr_2005. It was 145th for some time.

References

Agrawal, A., Yadama, G., Andrade, R. and Bhattacharya, A. (2006) 'Decentralization and environmental conservation: Gender effects from participation in joint forest management', CAPRi Working Paper No 53, CAPRi, Washington, DC

Anau, N., Iwan, R., van Heist, M., Limberg, G., Sudana, M. and Wollenberg, E. (2005) 'Negotiating more than boundaries: Conflict, power and agreement building in the demarcation of village borders in Malinau', in C. J. P. Colfer (ed) *The Equitable Forest: Diversity, Community and Resource Management*, Resources for the Future/CIFOR, Washington, DC

Anonymous (2005) 'Hutan Rakyat Indonesia Sangat Prospektif untuk Industri Kehutanan' ['The people of Indonesia truly look ahead regarding forestry industries'], *Siaran PERS* S.375/11/PIK-1/2005

Barr, C., Resosudarmo, I. A. P., Dermawan, A., McCarthy, J. F., Moeliono, M. and Setiono, B. (2006) *Decentralization of Forest Administration in Indonesia: Implications for Forest Sustainability, Economic Development and Community Livelihoods*, Center for International Forestry Research, Bogor, Indonesia

Barr, C., Wollenberg, E., Limberg, G., Anau, N., Iwan, R., Sudana, I. M., Moeliono, M. and Djogo, T. (2001) *The Impacts of Decentralisation on Forests and Forest-Dependent Communities in Malinau District, East Kalimantan*, CIFOR, Bogor, Indonesia

Baviskar, A. (2005) 'Between micro-politics and administrative imperatives: Decentralisation and the watershed mission in Madhya Pradesh, India', in J. C. Ribot and A. M. Larson (eds) *Democratic Decentralisation through a Natural Resource Lens: Cases from Africa, Asia and Latin America*, Routledge, London

Bennett, C. P. A. (2002) 'Responsibility, accountability, and national unity in village governance', in C. J. P. Colfer and I. A. P. Resosudarmo (eds) *Which Way Forward? People, Forests and Policymaking in Indonesia*, Resources for the Future/Center for International Forestry Research (CIFOR), Washington, DC

Bun, Y. A. (2006) 'Governance and community-based forestry in Papua New Guinea', paper presented at the Yogyakarta Workshop on Forest Governance and Decentralization (4–6 September), Yogyakarta, Indonesia

Burns, D., Hambleton, R. and Hogget, P. (1994) *The Politics of Decentralisation: Revitalising Local Democracy*, Macmillan Publications, Basingstoke, UK

Capistrano, D. and Colfer, C. J. P. (2005) 'Decentralization: Issues, lessons and reflections', in C. J. P. Colfer and D. Capistrano (eds) *The Politics of Decentralization: Forests, Power and People*, Earthscan, London

Casson, A. (2001a) 'Decentralisation of policies affecting forests and estate crops in Kotawaringin Timur District, Central Kalimantan', CIFOR, Bogor, Indonesia

Casson, A. (2001b) 'Decentralisation of policies affecting forests and estate crops in Kutai Barat District, East Kalimantan', CIFOR, Bogor, Indonesia

Colfer, C. J. P. and Capistrano, D. (eds) (2005) *The Politics of Decentralization: Forests, Power and People*, Earthscan, London

Contreras, A. P. (2003a) 'Creating space for local forest management: The case of the Philippines', in D. Edmunds and E. Wollenberg (eds) *Local Forest Management*, Earthscan, London

Contreras, A. P. (ed) (2003b) *Creating Space for Local Forest Management in the Philippines*, La Salle Institute of Governance, Manila, Philippines

Dahal, G. R. (2006) 'Governance and institutional structure: An ignored dimension of devolution policy process in forestry (the case of community based forest management in the Philippines)', PhD thesis, International and Rural Development Department, University of Reading, Reading, UK

Dangol, S. (2005) 'Participation in decision-making in Nepal', in C. J. P. Colfer (ed) *The Equitable Forest: Diversity, Community and Resource Management*, Resources for the Future/CIFOR, Washington, DC, pp54–71

Departemen Kehutanan, F. D. (2005) *Prosiding Workshop Penguatan Desentralisasi Sektor Kehutanan di Indonesia [Proceedings of the Workshop on Strengthening Forestry Sector Decentralization in Indonesia]*, Workshop Penguatan Desentralisasi Sektor Kehutanan di Indonesia, Pusat Rencana Kehutanan, Badan Planologi, Jakarta

Diaw, M. C. and Kusumanto, T. (2005) 'Scientists in social encounters: The case for an engaged practice of science', in C. J. P. Colfer (ed) *The Equitable Forest: Diversity, Community and Resource Management*, Resources for the Future/CIFOR, Washington, DC

Edmunds, D. and Wollenberg, E. (eds) (2003) *Local Forest Management*, Earthscan, London

Enters, T., Durst, P. B. and Victor, M. (eds) (2000) 'Decentralization and devolution of forest management in Asia and the Pacific', RECOFTC Report No 18 and RAP Publication 2000, Bangkok, Thailand

Gregersen, H., Contreras-Hermosilla, A., White, A. and Phillips, L. (2004) *Forest Governance in Federal Systems: An Overview of Experiences and Implications for Decentralization: Work in Progress*, CIFOR, Bogor, Indonesia

Hartanto, H., Lorenzo, M. C. B., Valmores, C., Arda-Minas, L., Burton, L. and Prabhu, R. (2003) *Learning Together: Responding to Change and Complexity to Improve Community Forests in the Philippines*, CIFOR, Bogor, Indonesia

Kaimowitz, D., Flores, G., Johnson, J., Pacheco, P., Pavez, I., Roper, J. M., Vallejos, C. and Velez, R. (2000) *Local Government and Biodiversity Conservation: A Case from the Bolivian Lowlands. A Case Study for Shifting the Power: Decentralization and Biodiversity Conservation*, Biodiversity Support Program, Washington, DC

Kartodihardjo, H. (2002) 'Structural problems in implementing new forestry policies', in C. J. P. Colfer and I. A. P. Resosudarmo (eds) *Which Way Forward? People, Forests and Policymaking in Indonesia*, Resources for the Future/CIFOR, Washington, DC

Komarudin, H., Siagian, Y. and Colfer, C. J. P. with Neldysavrino, Yentirizal, Syamsuddin and Deddy Irawan (2007) 'Collective action to secure property rights for the poor: A case study in Jambi Province, Indonesia', report to CAPRi (Collective Action and Property Rights System-wide Initiative), Bogor, Indonesia

Larson, A. M. (2005) 'Formal decentralisation and the imperative of decentralisation "from below": A case study of natural resource management in Nicaragua', in J. C. Ribot and A. M. Larson (eds) *Democratic Decentralisation through a Natural Resource Lens: Cases from Africa, Asia and Latin America*, Routledge, London

Larson, A. M. and Ribot, J. C. (2005) 'Democratic decentralisation through a natural resource lens: An introduction', in J. C. Ribot and A. M. Larson (eds) *Democratic Decentralisation through a Natural Resource Lens: Cases from Africa, Asia and Latin America*, Routledge, London

Lindayati, R. (2002) 'Ideas and institutions in social forestry policy', in C. J. P. Colfer and I. A. P. Resosudarmo (eds) *Which Way Forward? People, Forests and Policymaking in Indonesia*, Resources for the Future/CIFOR, Washington, DC

Liu, D. and Edmunds, D. (2003) 'The promises and limitations of devolution and local forest management in China', in D. Edmunds and E. Wollenberg (eds) *Local Forest Management*, Earthscan, London

Lynch, O. J. and Harwell, E. (2002) *Whose Natural Resources? Whose Common Good? Towards a New Paradigm of Environmental Justice and the National Interest in Indonesia*, Lembaga Studi dan Advokasi Masyarakat (ELSAM), Jakarta

Makapukaw, A. L. S. A. D. and Mirasol, F. S. (2005) 'Decentralizing protected area management at Mount Kitanglad', in C. J. P. Colfer and D. Capistrano (eds) *The Politics of Decentralization: Forests, Power and People*, Earthscan, London

Malysheva, N. V. (2005) 'Main features of Russia's forest management system', in C. J. P. Colfer and D. Capistrano (eds) *The Politics of Decentralization: Forests, Power and People*, Earthscan, London

Marzoni (2006) 'Negosiasi batas wilayah desa' ['Negotiating village territory boundaries'], in E. L. Yuliani, D. Tadjudin, Y. Indriatmoko, D. Munggoro, F. Gaban, and F. Maulana (eds) *Kehutanan Multipihak: Jalan Menuju Perubahan* [*MultiStakeholder Forestry: The Road to Change*], CIFOR, Bogor, Indonesia

McCarthy, J. F. (2001a) 'Decentralisation and forest management in Kapuas District, Central Kalimantan', CIFOR, Bogor, Indonesia

McCarthy, J. F. (2001b) 'Decentralisation, local communities, and forest management in Barito Selatan District, Central Kalimantan', CIFOR, Bogor, Indonesia

McGrath, S., McCarthy, J. and Moeliono, M. (2005) 'Forest and land tenure, resource-use decision making and spatial planning, synthesis report, Theme 2, Can decentralization work for forests and the poor?', CIFOR, DFID and ACIAR, Bogor, Indonesia

Mearns, R. (2005) 'Decentralisation, rural livelihoods and pasture-land management in post-socialist Mongolia', in J. C. Ribot and A. M. Larson (eds) *Democratic Decentralisation through a Natural Resource Lens: Cases from Africa, Asia and Latin America*, Routledge, London

Munggoro, D. W., Yuliani, E. L. and Indriatmoko, Y. (2006) 'The people of Baru Pelepat: Learning to change', video CD, CIFOR, Bogor, Indonesia

Ngakan, P. O., Achmad, A., Wiliam, D., Lahae, K. and Tako, A. (2005) 'The dynamics of decentralization in the forestry sector in South Sulawesi: The history, realities and challenges of decentralized governance', CIFOR, Bogor, Indonesia

Obidzinski, K. and Barr, C. (2003) 'The effects of decentralisation on forests and forest industries in Berau District, East Kalimantan', CIFOR, Bogor, Indonesia

Oyono, P. R. (forthcoming) 'Organizational infrastructure of local forest management in Cameroon: Elements of the sociology of natural resources for social theory and public policy', WRI/CIFOR, Yaounde, Cameroon

Pacheco, P. (2002) 'The implications of decentralization in forest management: Municipalities and local forest users in lowland Bolivia', World Resources Institute Conference on Decentralization and the Environment, Bellagio, Italy

Permatasari, E. (2007) 'Kembali ke Ladang Menggapai Asa' ['Back to the fields, seizing fate'], in Y. Indriatmoko, E. L. Yuliani, Y. Tarigan, F. Gaban, F. Maulana, D. Munggoro, D. Lopulalan and H. Adnan (eds) *Dari Desa ke Desa: Dinamika Gender dan Pengelolaan Kekayaan Alam* [*From Village to Village: Gender Dynamics and the Management of Nature's Wealth*], CIFOR, Bogor, Indonesia

Potter, L. and Badcock, S. (2001) 'The effects of Indonesia's decentralisation on forests and estate crops: Case study of Riau Province, the original districts of Kampar and Indragiri Hulu', CIFOR, Bogor, Indonesia

Resosudarmo, I. A. P. (2005) 'Closer to people and trees: Will decentralisation work for the people and the forests of Indonesia?', in J. C. Ribot and A. M. Larson (eds) *Democratic Decentralisation through a Natural Resource Lens: Cases from Africa, Asia and Latin America*, Routledge, London

Resosudarmo, I. A. P. and Dermawan, A. (2002) 'Forests and regional autonomy: The challenge of sharing the profits and pains', in C. J. P. Colfer and I. A. P. Resosudarmo (eds) *Which Way Forward? People, Forests and Policymaking in Indonesia*, Resources for the Future/CIFOR, Washington, DC

Resosudarmo, I. A. P., Barr, C., Dermawan, A. and McCarthy, J. (2006) 'Fiscal balancing and the redistribution of forest revenues', in C. Barr, I. A. P. Resosudarmo, A. Dermawan, J. F. McCarthy, M. Moeliono and B. Setiono (eds) *Decentralization of Forest Administration in Indonesia*, CIFOR, Bogor, Indonesia

Ribot, J. C. and Larson, A. M. (eds) (2005) *Democratic Decentralisation through a Natural Resource Lens: Cases from Africa, Asia and Latin America*, Routledge, London

Samsu, Suramenggala, I., Komarudin, H. and Ngau, Y. (2005) 'The impacts of forestry decentralization on district finances, local community and spatial planning: A case study in Bulungan District, East Kalimantan', CIFOR, Bogor, Indonesia

Sarin, M., Singh, N. M., Sundar, N. and Bhogal, R. K. (2003) 'Devolution as a threat to democratic decision-making in forestry? Findings from three states in India', in D. Edmunds and E. Wollenberg (eds) *Local Forest Management*, Earthscan, London

Siagian, Y., Morgan, B., Yentirizal and Neldysavrino (2006) 'Strengthening women's participation through collective action for inclusive decision-making processes: Lessons learned from two villages in Jambi Province, West Sumatera, Indonesia', paper presented at 'Survival of the Commons: Mounting Challenges and New Realities', the Eleventh Conference of the International Association for the Study of Common Property, Bali, Indonesia, 19–23 June

Siswanto, W. and Wardojo, W. (2005) 'Decentralization of the forestry sector: Indonesia's experience', in C. J. P. Colfer and D. Capistrano (eds) *The Politics of Decentralization: Forests, Power and People*, Earthscan, London

Sithole, B. (2005) '"Becoming men in our dresses!" Women's involvement in a joint forestry management project in Zimbabwe', in C. J. P. Colfer (ed) *The Equitable Forest: Diversity, Community and Resource Management*, Resources for the Future/CIFOR, Washington, DC

Soetarto, E., Sitorus, M. T. F. and Napiri, M. Y. (2003) 'Decentralisation of administration, policy making and forest management in Ketapang District, West Kalimantan', CIFOR, Bogor, Indonesia

Sudirman, Wiliam, D. and Herlina, N. (2005) 'Local policy-making mechanisms: Processes, implementation and impacts of the decentalized forest management system in Tanjung Jabung Barat District, Jambi', CIFOR, Bogor, Indonesia

Tokede, M. J., Wiliam, D., Widodo, Gandhi, Y., Imburi, C., Patriahadi, Marwa, J. and Yufuai, M. C. (2005) 'The impact of special autonomy in Papua's forestry sector: Empowering customary communities (*Masyarakat Adat*) in decentralized forestry development in Manokwari District', CIFOR, Bogor, Indonesia

Turner, M. and Hulme, D. (1997) *Governance, Administration and Development: Making the State Work*, Macmillan Publications, Basingstoke, UK

Vernooy, R. (ed) (2006) *Social and Gender Analysis in Natural Resource Management: Learning Studies and Lessons from Asia*, Sage Publications, London

Véron, R., Williams, G., Corbridge, S. and Srivastava, M. (2006) 'Decentralized corruption or corrupt decentralization? Community monitoring of poverty-alleviation schemes in Eastern India', *World Development*, vol 14, no 11, pp1922–1941

Wiliam-de Vries, D. (2006) *Gender Bukan Tabu: Catatan Perjalanan Fasilitasi Kelompok Perempuan di Jambi* [*Gender is not Taboo: Notes on the Process of Facilitation of Women's Groups in Jambi*], CIFOR, Bogor, Indonesia

Wiliam-de Vries, D. and Sutarti, N. (2006) 'Gender equity: Revealing the reality for the women of Jambi', CIFOR Governance Brief 29, CIFOR, Bogor, Indonesia

Wollenberg, E., Anderson, J. and Edmunds, D. (2001) 'Pluralism and the less powerful: Accommodating multiple interests in local forest management', *International Journal of Agricultural Resources, Governance and Ecology*, vol 1, nos 3/4, pp199–222

Wrangham, R. (2002) 'Changing policy discourses and traditional communities, 1960–1999', in C. J. P. Colfer and I. A. P. Resosudarmo (eds) *Which Way Forward? People, Forests and Policymaking in Indonesia*, Resources for the Future/CIFOR, Washington, DC

Yasmi, Y., Anshari, G. Z., Alqadrie, S., Budiarto, T., Ngusmanto, Abidin, E., Komarudin, H., McGrath, S., Zulkifli and Afifudin (2005) 'The complexities of managing forest resources in post-decentralization Indonesia: A case study from Sintang District, West Kalimantan', CIFOR, Bogor, Indonesia

Yuliani, E. L., Ernawati, S., Indriatmoko, Y. and Colfer, C. J. P. (2006) 'Strengthening women's roles (*Penguatan Peran Perempuan*)', poster presented in CBFM Festival, CIFOR, Jakarta, Indonesia

Annex: The Yogyakarta Workshop and Associated Field Trips

Ganga Ram Dahal and Yunety Tarigan

The workshop took place from 4–6 September 2006 in Yogyakarta, Indonesia, with over 175 participants from 25 countries, with about 40 per cent of the participants Indonesian. Countries represented included Australia, Bangladesh, Bhutan, Cambodia, China, India, Indonesia, Japan, Korea, Lao PDR, Madagascar, Malaysia, Mongolia, Myanmar, Nepal, New Zealand, Pakistan, Papua New Guinea, the Philippines, Russia, Sri Lanka, Switzerland, Thailand, Timor Leste, the UK, the US and Vietnam.

There were three main substantive phases that corresponded with the full days of the workshop. The first day was devoted to presentations, beginning with two overview presentations in plenary, after which the groups divided into three parallel sessions, dealing with: a) roles, rights and responsibilities; b) institutions, including organization, capacity-building and legal aspects; and c) financial implications of decentralization. The second day was devoted to field trips. On the third day, the group broke into five working groups to discuss the workshop themes mentioned above. To understand local realities about forestry sector decentralization in Central Java, the participants of the Yogyakarta workshop visited the following sites, each of which had its own scheme.

Site 1: Social forestry in Tani Manunggal, Playen sub-district, Gunung Kidul.

In Tani Manunggal, local communities formed a cooperative to manage 40ha of forest land in 2003. The legal basis under which the community forest scheme functions is *Bupati* (district head) Decree No 213/2003. This decree, valid for five years, took effect after decentralization in 2001, when social forestry schemes became the responsibility of the districts. Forming a cooperative is obligatory. Local communities have planted teak, mahogany and eucalyptus to rehabilitate the degraded lands, along with simultaneous intercropping of food crops (the practice is widely known as *tumpangsari*) for subsistence needs.

Site 2: Forest land rehabilitation in Paliyan, Gunung Kidul.

At present, two- and three-year-old teak and mahogany trees cover 305ha in Kepek, Paliyan. Approximately 707 households are involved in the rehabilitation scheme, plus the Ministry of Forestry, Forestry and Plantation Services of Yogyakarta, the Watershed Management Institute, and local communities. Local farmers are growing cassava, maize and papaya inside the plantation area to support their livelihoods. Finances for the scheme are controlled by the central government but pass through the Provincial Forest Service. In the long run, farmers plan to grow grasses and shift from food crops to livestock.

Site 3: Private forest activities in Sumberrejo and Selopuro villages in Wonogiri District.

The local communities initiated the planting of trees which transformed the landscape into a green site with impressive teak and mahogany trees on some 812ha in the two villages. With the help of a local NGO and district forest offices, local communities had their forest certified by Lembaga Ekolabel Indonesia for sustainable community-based forest management. As a result of the plantation, the watershed was restored and water is now available to irrigate 40ha of paddy field in the villages.

Site 4: Community-based forest management for enterprises in Kedu Selatan, Kaligesing sub-district.

The community of Kedu Selatan is managing 170ha of forest land under a state-owned forest enterprise (*Perum Perhutani*) scheme, under which community members and the government share benefits. The scheme operates under a decree of the governor of Central Java, No 24/2001. Community members are organized in a group for the promotion of forest-based enterprises. Besides planting trees in the specified area, they are allowed to grow food crops among the trees, tap resin and gum from the pine trees, and collect leaves for their livestock.

Based on observations during the field visits in Central Java, it is clear that decentralization in Indonesia is at an early stage and many issues remain unresolved. Probably the most important tasks remaining are to clarify the roles, responsibilities and authorities of different actors; support local institutions to develop their human and financial capacity; and win the confidence of local communities through establishment of a clear benefit-sharing scheme, making sure that the tenure and use rights of local people will continue for the long run.

2

From Decentralization to Governance: Recurring Issues

Christian Küchli

Decentralization and governance are themes to which Switzerland has traditionally attributed high importance in international cooperation. As with many other countries, Switzerland believes that decentralized forest governance is pivotal in addressing most of the critical issues related to sustainable forest management. At meetings of the Committee on Forestry of the Food and Agriculture Organization in Rome in 2005, for example, many participants stated that community-based forestry was particularly well placed to address poverty reduction. It was repeatedly mentioned that much remained to be done, since the underlying causes of deforestation, such as land tenure uncertainty and inequity, had yet to be addressed.

Similarly, in fire management, case studies have shown that community forestry practices have reduced damaging and unwanted fires, and that increased participation and access to forest ownership have led to more effective fire prevention and suppression.

In many other areas as well, decentralization is perceived to be part of the solution. In the field of forest law enforcement and governance, Switzerland is focusing on governance issues as central to overcoming the problem of illegal logging. In the drive to reduce illegal logging, forest-dependent communities can become best allies, given the correct incentives. The same is true for forest rehabilitation: addressing the underlying factors, such as land tenure and access to forest resources, encourages benign and beneficial local participation.

A further reason to support forest governance as a key to sustainable forest management is the fact that many problems we see in Asia and the Pacific today have also appeared in Switzerland's forest history. In fact, forest governance all over the world may share a similar set of fundamental problems.

Although there are certainly no standardized solutions to address the diversity of problems that exist in the world's forests, a short historical journey to Switzerland may help identify basic problems, potential solutions and the possible timeframe for fundamental change.

The Swiss Case

Switzerland consists of 26 cantons and – very important from the point of view of forest governance and sustainable forest management – some 3000 communes and even more communities. Rural areas in Switzerland traditionally have a high level of local autonomy. For many generations, local communities, often several in one political commune, have administered common lands like forests and have acted as custodians of forest-related knowledge. Unlike regions with a colonial history, local autonomy in Switzerland has never been disrupted.

Nevertheless, despite strong local autonomy stretching back many centuries, property rights to forests in rural areas have been claimed by the powerful cities in the lower elevations of Switzerland. In general, only use rights to forests have been conceded to local communes. This legal framework caused latent conflicts between cities and rural areas. For example, when the City of Berne sometimes sent timber concessionaires to Alpine regions to float fuelwood for the winter to the city, the local population would react with anger. One of the consequences of such governmental interventions was uncertainty about use rights and property at the local level, and therefore a less careful management of the forest by the rural population.

It was only with the liberal winds of change in the 1830s that forest property ownership was decentralized. The cantons started to hand property rights over to the communes. Initially, decentralization had disastrous impacts on the forests. Immediately after the reallocation of forest property, timber exploitation became highly speculative. In a climate of legal uncertainty, people cut whatever they could, especially where boundaries between adjacent owners had not yet been established. Latent disputes flared into conflicts. Powerful market forces hastened the deforestation: Europe's growing economies generated high timber demand, and considerable volumes of wood were sold for almost nothing to timber merchants from France or Holland.

The few foresters of the time held no illusions concerning the enforcement of the existing legal measures now that the gates to timber export had been opened wide. In the 1820s, one of them noted that forestry regulations 'cannot be enforced in Alpine cantons only by public officials and the forest service. Legions of officials would be required' (Kasthofer, 1828, p123) to defend the forest against actions of the local population.

In the 1860s, the large-scale clear-cutting of Alpine forests contributed to extensive flood disasters, affecting even the cities in the river valleys. This was the moment when the federal government intervened in the forestry sector, eventually leading to the enactment of the first Federal Forestry Law in 1876.

The federal law had to accommodate the interests of the Alpine cantons. To achieve that, it was designed within a federal framework with four overarching regulations:

1 forest area could no longer be reduced in size;
2 owners of public forests were required to clearly specify proposed timber harvesting in a forest management plan;
3 areas where timber harvesting took place had to be replanted within three years; and
4 significantly, the Swiss confederation contributed to the salaries of the cantonal foresters.

At the second legislative layer, the forest laws of the 26 cantons had to comply with the federal law. The third legislative layer, the regulations established at the local level, addressed the design of transparent community institutions for making management decisions, distributing benefits, resolving conflicts, and defending community interests against the exploitive outside forces and local elites.

Local capacities as well as decision-making power and accountability at the local level, combined with the strong guiding provisions at the federal and cantonal level, proved critical. The forests recovered, and at the beginning of the 20th century, sustainable forest management and timber production were a significant stimulus to local economies and also satisfied national and international demand for products and services, especially coming from protective forests (i.e. forest that protects against avalanches, rockfall, etc.).

Universal Issues

Several of the issues that emerged in the Interlaken Workshop (Chapter 1) and its 1998 antecedent, the Seminar on Decentralization and Devolution of Forest Management in Asia and the Pacific, held in Davao in the Philippines, echo the forest management challenges Switzerland has faced.

Transfer of authority and the role of the forest service

The first shortcoming that the workshops in Davao and Interlaken identified is that the transfer of significant authority over forest resources to local governments is rare, and that in many cases no or only few discretionary powers are handed over. When powers are granted, it is typically only over a small area, or over resources with little economic value. Often, central governments were reported to block decentralization or to manipulate it to their own ends. Where decentralization takes place, it can be an imperfect measure: powers transferred by central governments get stuck at sub-national levels and are not further transferred down to the local level.

One possible explanation for this reluctance, especially of central governments, could be the prevailing focus on decentralization per se, rather than on a more comprehensive view of forest sector governance. If major actors perceive decentralization as giving something away or losing something, or if they consider decentralization a threat to their professional standing, their motivation is likely to be limited. Decentralization in Switzerland was also about changing power relations over an economically important resource, and given the high value of tropical trees, change will be especially difficult in Asia and the Pacific. Forest departments are often directly financed through timber concessions and stumpage fees. In decentralized situations such resources tend to decrease or run dry.

A change of forest governance toward a more decentralized system will succeed only if such political and economic constraints and the resulting institutional desire to avoid change are surmounted. Preconditions to achieve this are that foresters assume a new role in forest governance. In Switzerland, the salaries of the cantonal forest service have been subsidized by the Swiss confederation, and today more than ever, salaries that secure a reasonable living are certainly determining factors in the agency's morale and effectiveness.

The more centrally forests are managed, the more controlling and policing functions foresters have. If experience in the Alps is any guide, foresters in newly decentralized situations will remain important and have enhanced status. Before decentralization, Swiss foresters were policemen who enforced provisions in the interests of powerful cities or cantonal governments. This was the time, as a popular writer remarked, when forest wardens knew as much about silviculture as a hen knows about playing the violin (cf. Küchli and Stuber, 2000). After decentralization they became facilitators and forest management consultants to local forest owners. A first step, establishing borders between communes and resolving demarcation disputes, took decades in Switzerland – 50 years in one case. Foresters' scientific knowledge combined well with local forest-related knowledge, and thus the cultural and biological realities of the different geographical areas were accommodated.

Governance: From command-and-control to interaction

Decentralization is not an end in itself but an element of the overlying concept of governance – a necessary first step in establishing forest sector governance that works. The Davao and Interlaken workshops focused on decentralization because there is an urgent need for action to make forest governance work better or, in some cases, start to work at all. But in a comprehensive view of forest sector governance, all actors are important, not just the local ones. Decentralization and centralization are two sides of the same coin: it is the interaction within a governance system that ultimately determines the balance and fate of the forest resources in all their aspects – social, economic and ecological.

The next step should be a focus on the interplay among all the actors. Governance embraces the interactions among different levels – local, sub-national and national – as well as across sectors, and, finally, regionally and globally. Working forest governance systems have a clear set of rules on how to reach consensus and make decisions, including the mechanisms to hold decision-makers accountable for their actions.

Working forest governance uses not rules that cannot be enforced, but rather incentives. In Switzerland, responsibilities like watershed management were imposed but accompanied by funding. The confederation provided compensation as an incentive for reforestation of protective forests. This was the beginning of Switzerland's cooperative federalism with 'coupled contributions': the confederation and the cantons share the costs of projects in protective forests.

Capacity-building to allow conflict resolution

Capacity-building, a crucial element of successful decentralization, is a process of empowerment at different scales. The local level – local government units, communes, communities, and user groups – needs support for this endeavour the most. Necessary skills are articulation of local interests, negotiation and conflict management, and financial accounting, especially with respect to benefit sharing.

Inequitable sharing of benefits was and is a main cause of conflicts, the reason often being 'elite capture'. Mechanisms to reduce this include developing policies that strengthen checks and balances as well as bringing equity and decision-making issues into public discussion at all levels. Equitable benefit sharing requires institutions that are representative of and are accountable to local populations, accountability being one of the indispensable features of decentralized governance systems. Security of resource access and the stability of the legal environment are also essential to successful decentralization. There is in any case a need to establish a fair and transparent process with the participation and consent of local communities, and a reliable tax system.

Collective local control or title can be more secure and more effective in inducing sustainable forest management over the long run than titles held by individuals. Whenever possible, traditional communal forest management systems with silvicultural practices that favour multiple functions should provide the basis for decentralization. The people who live in and around forests have considerable knowledge of forest ecology and especially biodiversity. Reliance on local knowledge and links with local cultures can make decentralized systems more resilient in the face of changing external conditions.

Transition time

Decentralization, by breeding chaos and enabling illegal logging, has been blamed for the destruction of forests. Such outcomes have been observed in some places during transition periods, especially if the transition is abrupt and not well prepared. In Switzerland, too, decentralization initially had disastrous impacts on the forest due to the high timber demand and the uncertainty triggered by the transfer of property rights.

The transition to decentralized decision-making and management of forests is likely to be costly, and the pay-offs may not materialize immediately. During the transition phase, central government funds are needed for strengthening capacity and providing guidance to local governments and stakeholders. Once consolidated, with a robust framework of tenure rights, conflict management and effective incentives, decentralized systems are better fitted to regain and to maintain stability and will often favour more biologically diverse and locally adapted approaches to restoration.

Decentralized governance systems are complex and dynamic processes that need constant learning; this requires time and long-term commitment. Sustainable forest management is less about trees than about people and organizations and the way they interact. This is why sustainable forest management may take even longer to grow than the trees themselves.

References

Küchli, C. and Stuber, M. (2000) 'Forest development in the Swiss Alps: Exchanging experience in the Swiss Alps with mountain regions in the south', CD-ROM, Swiss Agency for Development and Cooperation (SDC), Bern

Kasthofer, K. (1828) *Der Lehrer im Walde, Teile I und II* [*The Teacher in the Forest, Part I and II*], Jenni, Bern

3

Governance, Tenure and Equity in Asia-Pacific Forests

Lesley Potter

An important recent paradigm shift in forest development in Asia and the Pacific (as in many parts of the world) has been the move from centralized to decentralized forest management. If, however, governance is defined as 'the procedures and rules by which decisions are made and consensus is reached, as well as mechanisms to hold decision makers accountable for their actions' (Menzies, 2004, p450), the extent to which powers of decision-making and enforcement are really being devolved to forest communities remains questionable.

'Forest-rich' countries such as Indonesia have been especially slow to devolve such powers. Where local communities have been given the responsibility for restoration of degraded areas, the products from regenerated forests have frequently been appropriated or taxed. Questions of gender, class and caste continue to divide communities: decentralization does not by itself ensure equality and may pose greater challenges for women as traditional values are revived. Policies of decentralization and devolution that do not include adequate community participation or genuine downward shifts of power and responsibility cannot produce favourable outcomes in terms of poverty alleviation or sustainable forest management.

Community perception of tenure security is vital (though formal title is not always essential) and affects such problems as forest fire and some forms of illegal logging. Tenure security includes not only effective community institutions but also the ability to respond to critical turning points when political conditions change. In many countries tenure security has been difficult to achieve, as governments continue to claim ownership of the forests and traditional tenure is unrecognized. This situation leaves local farmers vulnerable to pressures from both government and corporate interests. Paradoxically, in Papua New Guinea, where forests are owned by local people, the government has implemented a forest policy based on export logging without adequate consultation or empowerment (Bun et al, 2004; Turia, 2005; Box 3.1).

This overview examines these issues, beginning with the state of the forests as measured by the FAO (2005a), followed by a general discussion of decentralization and devolution. The major findings are identified from two earlier conferences (Davao in 1998 and Interlaken in 2004) together with conclusions from other relevant sources. Such findings formed the background to the 2006 meeting in Yogyakarta.

Box 3.1 Governance and community-based forestry in Papua New Guinea

Lesley Potter and Yati A. Bun

A stated objective of the Papua New Guinea Forestry Act of 1991 was to maximize local participation in the development of forest resources as a renewable asset. However, since the beginning of export logging in 1979, commercial extraction has been almost entirely controlled by foreign companies.

Forest governance and decentralization

Each of the 19 provinces has a forest management committee which develops forestry plans. These are passed to the National Forestry Board and eventually the Minister of Forests for endorsement. After offering a timber area for sale, the Board is supposed to consult the customary owners and the provincial government. Landowners form a company, obtain a timber permit and then sign a 'logging and marketing agreement' with a foreign logging firm. This approach has failed: in October 2006, there was no successful landowner company operating at any logging concession in the country.

Community forestry in Madang

The Madang Forest Resource Owners Association (MFROA) represents an effort to make community forestry successful and sustainable. Those wanting to develop their own forest resources seek help from the Foundation for People and Community Development Inc, a local NGO, to get them started. The NGO provides training in forest and business management, portable sawmill technology and maintenance. MFROA members will soon be able to market fully certified eco-timber. This is a significant milestone as no large-scale commercial forestry operations in Papua New Guinea have earned such independent certification.

The experience of Madang exemplifies the kinds of problems and opportunities that exist in connection with community forestry. In the 1970s Madang was the first province to experience large-scale clear-felling. The unfavourable environmental and social impacts persuaded many landowners to place higher value on their forests and seek to manage them themselves. However, the use of their assets is seriously restricted by government regulations and lack of proper assistance. Obtaining the necessary permit for cutting timber involves 34 steps, and compliance with this regulation is onerous. Most villagers are illiterate and fall easy victims to unscrupulous negotiators. Local landowners face four particularly difficult problems:

1 Not knowing their rights: Landowners are not given enough information to understand their rights. They have a right not to sign over their forest resources: it should be made clear that they have a choice.
2 Not knowing how to begin: Some do not know that they need a licence to cut timber for commercial purposes, even when they own the forest land.
3 Lacking capital and technical support: Many landowners wish to develop and manage their forest resources but lack capital and need regular extension services. Community forestry is new and not well understood by government. It does not generate as much government revenue as industrial logging.
4 Making a commitment to hard work: Managing one's own forest resources is not easy. Getting agreement from all members of a clan is a major challenge, with members often needing to pool resources. Many questions arise regarding operations, marketing and income distribution.

Changes needed on the part of government to assist community forestry

- Current forestry laws primarily support round log exports and large-scale logging activities, and are too cumbersome for community forestry.
- The provincial forest management committees do not function well. They are critical to bring decision-making in forestry closer to the grass roots.
- The development of forestry resources is a major investment in bringing government services to rural dwellers. Special programmes and adequate funding are needed for communities to take up forest management.
- The potential for harvesting and marketing non-timber forest products, which could improve local economies, is inadequately understood.

Conclusion

In Papua New Guinea, land and forest are customarily owned. The rightful landowners need to be given the opportunity to manage their forest resources themselves with the encouragement and support of the national government. The rights of customary landowners should not be removed.

The State of Forests in the Asia-Pacific Region

The countries of the Asia-Pacific region[1] cover almost 22 per cent of the world's land area and account for 19 per cent of its forests. They are also home to more than 55 per cent of the world's population, though both forests and people are unevenly distributed (FAO, 2005a). The current state of the region's forests provides the backdrop to the analysis that follows.

The most recent global definition of forests identifies 'trees higher than 5 metres and a canopy cover of more than 10 per cent, or trees able to reach these thresholds in situ' (FAO, 2005a, p69). Regenerating forests are included, as (more controversially) are forest plantations, which are now assuming considerable importance. 'Primary' forests consist of native species 'with no clearly visible indications of human activities' (FAO, 2005a, p171). Local experts might dispute the figures, and others would question the continuing existence of primary forests. The value of the FAO numbers, however, is their broad comparability for the identification of general trends.

Several countries in Southeast Asia, especially Indonesia, Vietnam and Cambodia, together with Papua New Guinea in Oceania, experienced a high annual loss of their 'primary' forests between 2000 and 2005 (Table 3.1). In Northeast Asia, on the other hand, Japan actually increased its primary forests and retained a high proportion of its land under forest cover, though 41 per cent of that consisted of plantations. While Indonesia's primary forests declined by 1.4 million hectares per year, the area of plantations in China rose by virtually the same amount, 1.5 million hectares per year. Vietnam, India, South Korea, Indonesia and Myanmar were involved to varying degrees in new plantation development; similar planting occurred earlier in Thailand and the Philippines (FAO, 2005a, Annex 3). The larger countries of Oceania, such as Australia and New Zealand, also have important areas of plantations. Regeneration of natural forest has become a major concern in the rather forest-poor countries of South Asia (including India and Nepal) and in China and Vietnam. Much of this regeneration is now in local hands.

Within the region, one of the most important drivers of forest change has been the rapid emergence of China as a huge market for timber and wood products from Southeast Asia and Oceania (as well as Russia), replacing Japan, which had previously dominated the regional timber trade. Northern Myanmar, Indonesia (especially Papua) and Papua New Guinea have been major sources, with a high proportion of the timber being obtained illegally (Bun et al, 2004; EIA/Telapak, 2005; Global Witness, 2003, 2005; White et al, 2006). This new trade is partly responsible for some of the rapid recent declines in natural forests.[2]

Table 3.1 *Forests of Asia-Pacific nations, 2005*

	Forest area (000ha)	*Per cent of total land area*	*Percentage change, forest and woodlands 2000–2005*	*Primary forest as percentage of forest area*	*Change 2000–2005 (ha/yr)*	*Plantations as percentage of forest area*	*Change 2000–2005 (ha/yr)*
Northeast Asia							
China	197,290	21.2	+2.2	5.9	0	15.9	+1,489,000
Japan	24,868	68.2*	ns	18.5	+107,400	41.5	–2000
Rep. Korea	6265	63.5*	–0.1	ns	ns	21.8	+35,200
Southeast Asia							
Cambodia	10,447	59.2*	–2.0	3.1	–26,800	0.6	–2600
Indonesia	88,495	48.8*	–2.0	55.0	–1,447,800	3.8	+79,400
Lao PDR	16,142	69.9*	–0.5	9.2	ns	1.4	+25,000
Malaysia	20,890	63.6*	–0.7	18.3	0	7.5	–17,200
Myanmar	32,222	49.0*	–1.4	ns	ns	2.6	+30,600
Philippines	7162	24.0	–2.1	11.6	0	8.7	–46,400
Thailand	14,520	28.4	–0.4	44.4	0	21.3	+4400
Vietnam	12,931	39.7	+2.0	0.7	–20,400	20.8	+129,000
South Asia							
Bangladesh	871	6.7	–0.3	ns	ns	32.0	+600
India	67,701	22.8	ns	ns	ns	4.8	+84,200
Nepal	3636	25.4	–1.4	9.6	–7000	1.5	+200
Pakistan	1902	2.5	–2.1	ns	ns	16.7	+4400
Sri Lanka	1933	29.9	–1.5	8.6	–6000	10.1	–5141
Oceania, Pacific							
Australia	163,678	21.3	–0.1	3.2	0	1.1	+56,200
Fiji	1000	54.7*	ns	89.4	0	10.1	0
New Zealand	8309	31.0	+0.2	42.2	0	22.3	+16,600
Papua New Guinea	29,437	65.0*	–0.5	85.6	–250,000	0.3	+1980
Solomon Islands	2172	77.6*	–1.7	ns	ns	ns	ns

Notes: * = 'Forest-rich' country; ns = not stated.
Source: FAO (2005a, Annex 3: Global Tables).

Asia-Pacific Experience

Devolution of aspects of forest management from government forest departments to local communities has become a widespread phenomenon in Asia and the Pacific in the past 30 years, part of an inexorable global trend (White and Martin, 2002a). The tentative beginnings of 'social forestry' projects were noted in the Philippines, Thailand, Java and India in the 1970s, at the same time as village associations in Nepal's Middle Hills were given some responsibility for degraded forest. The nature and scope of these approaches differed, but the delegation of authority to communities was very limited[3] (Gibbs et al, 1990; Pragtong and Thomas, 1990; Stoney and Bratamihardja, 1990; Poffenberger, 1990, 2000; Gilmour, 2003). The past 15 years have seen the most marked changes, with many more governments involved. The communist states of China, Vietnam and Laos have allocated land directly to households, mainly for purposes of reforestation. India has embarked on a large programme of joint forest management that now covers about 25 per cent of all forests. It is a collaborative effort between communities and individual states, especially four states with considerable tribal populations (Milne, 2005). The Philippines offers upland forest leases to settler migrants and certificates of ancestral domain to indigenous minorities. The growth of forest user communities to manage regenerated forests in Nepal has become famous, covering 28 per cent of the potential community forest area, though the approach works better in the Middle Hills than in the forest-rich *terai* (Gautam et al, 2004; Adhikari et al, 2006).

Why has this process taken place? Quickening rates of tropical deforestation during the 1980s led many experts at multilateral agencies to reconsider the wisdom of industrial forestry and the capacity of state agencies to sustain natural forests.[4] Many national forest departments were also accused of corruption, supporting illegal logging and abusing the forests for political gain (White and Martin, 2002a). The World Bank began promoting community forestry as a new model of development assistance, though issues of tenure were rarely addressed and projects mainly emphasized capital and technical aid (Poffenberger, 2000). Government agencies themselves became more receptive to some form of devolution, since policies excluding people from forests were clearly failing and conflicts were increasing, creating a 'sense of crisis' (Menzies, 2002, p6). Dupar and Badenoch (2002) and Larson (2005) attribute the change to common goals of cost reduction and improved efficiency among state forest agencies, rather than any more sympathetic stance toward indigenous or local people, as suggested by White and Martin (2002a).

There has been considerable scepticism about the extent to which powers of decision-making and enforcement are really being devolved to communities, in addition to the more onerous responsibilities of restoration, protection and monitoring. Edmunds and Wollenberg (2003, p158) conclude that devolution policies represent '*a shift in the manner in which central governments control forest management, rather than a genuine shift in authority to the poorest forest users*'.[5] Community-based forest management in the Philippines, often presented as one of the more successful Asian examples, has been lambasted as 'a misnamed continuation of central control ... [that] has enabled self-interested bureaucracies to inhibit the sharing and transfer of power' (Gollin and Kho, 2002, quoted in Menzies, 2002, p14; see also Menzies, 2004). Guiang et al (Chapter 11 of this volume) note a growing 'disconnect' between policy and reality, with the resource rights of communities under devolved forest management being regularly suspended or cancelled by the forest agency.

Not participating in the trend towards devolved forest management until recently were the four territories that still retained well-stocked forests: Indonesia (outside Java), East Malaysia (Sabah and Sarawak), Cambodia and Myanmar. Indonesia, with the thoroughgoing decentralization that occurred so swiftly after the demise of the authoritarian Suharto regime in 1998, offers perhaps the most interesting possibilities, despite the slow acceptance of change by the central Forestry Department.

Findings from the two conferences on decentralization

The 1998 Davao meeting, organized by the Bangkok Regional Community Forestry Training Center (RECOFTC), brought together representatives from Southeast and South Asia, and also from China, New Zealand and West Africa. In the overview of themes and issues, Fisher (2000, pvi) defined decentralization as 'relocation of *administrative functions* away from a central location' and devolution as 'the relocation of *power* away from a central location'.[6] I have adopted these useful definitions in this chapter. Three basic approaches were identified:

1 Decentralization without devolution. The Indian Joint Forest Management schemes provided an example. The state forest departments set the objectives; communities had responsibilities and some benefits but little or no authority.
2 Decentralization of forest management to local government but not to local communities. The Philippines was suggested as an example, but local governments still lacked power.
3 Handing over of a significant amount of control to local communities or individuals. Such an outcome was said to be rare, with Nepal the exception, but even there contradictions existed between the rights of forest user groups and those of local government. As one of the Nepal case studies pointed out, decentralization of government administration had undermined pre-existing community forestry (Upreti and Shrestha, 2000, p139).

In a separate paper, Fisher (2000) provided a further analysis of four concepts: decentralization, devolution, power and participation. In discussing participation, he noted that access to valuable forest resources was rarely provided, with community forestry often applying only to degraded areas. Banerjee explored the prospects of forest reform in which forests were actually returned to the people, who were given both power and management responsibility. However, he was not optimistic that such policies would be introduced because 'in spite of the rhetoric . . . participation is dismal' (Banerjee, 2000, p46).

Devolution in New Zealand took a unique turn: the government was selling parts of the forests (though not the land) to both local and foreign investors and distributing the functions of the forestry department to other agencies, such as conservation and agriculture (Clarke, 2000; see also Box 3.2).

Box 3.2 Forest sector restructuring in New Zealand: The 'big-bang' approach to decentralization

Lesley Potter and Colin O'Loughlin

New Zealand's forests occupy 31 per cent of the country's land area. The indigenous forests, covering mainly hill and mountain lands, are managed primarily for conservation and environmental protection, while introduced plantations of *Pinus radiata*, on 7 per cent of the land, supply 99 per cent of total timber production.

By the 1960s the state-run New Zealand Forest Service (first established in 1919) had developed into a complex organization controlling all aspects of forest management, including research, silviculture, some sawmilling and all timber sales. In the mid-1980s it employed over 7000 people. Through the 1970s and 1980s there was growing dissatisfaction with the service's organization and performance, both from the green movement, which lobbied for the indigenous forest estate to be managed by a conservation organization, and from the auditor general, who deemed the service's accounting system to be satisfactory for managerial but not for commercial purposes. The Treasury proposed that the commercial activities be transferred to a separate state-owned enterprise.

The government decided in December 1985 to split the New Zealand Forest Service into a commercial corporation to administer wood production and wood processing, and a separate organization, the Ministry of Forestry, to undertake sectoral and regulatory functions and research. A new Department of Conservation was set up to manage conservation lands, including the indigenous forests. On 1 April 1987 the New Zealand Forest Service was disestablished and its functions transferred to the New Zealand Forestry Corporation, the Ministry of Forestry and the Department of Conservation.

The rapidity of the changes and the limited discussion of the intended restructuring caused great distress. Affected individuals, communities and whole districts dependent on forestry for their livelihoods had little time to plan for and adjust to the new conditions, as 3205 staff lost their jobs. The situation was exacerbated three years later, when the state forests were privatized and the New Zealand

Forestry Corporation disappeared. The Treasury and the Forestry Corporation developed a design for the pending sale of forests that did not threaten any future ownership of the lands by the indigenous Maori population. The purchaser bought the existing trees, buildings and other fixed assets, but the land itself remained in Crown ownership. It could be leased back to the purchaser for approved activities such as planting replacement crops. Foreign ownership increased from less than 2 per cent in the 1980s to approximately 48 per cent after 1996.

Economic outcomes of privatization

The sale of the state's plantation forests between 1990 and 1996 realized NZ$3.4 million. Over the 1990s, total investment by foreign and domestic forestry companies, including investment in wood-processing facilities, approximated NZ$800 million. Between 1993 and 2000, 430,000ha were added to the plantation estate, most planted by farm foresters, investment foresters and other small landowners. After 2000, planting rates plummeted because of falling profitability.

Social implications of privatization

The social impacts of dismantling the New Zealand Forest Service and privatizing the plantations were widespread. Forestry towns suffered substantial hardship when large numbers of workers lost their jobs. In the Northland region, with a large Maori population, the changes were accompanied by a drastic drop in forestry labour requirements and a marked deterioration in the local economy.

CONCLUSIONS

- In designing large-scale restructuring, government needs to keep the public and the industry informed about what is being planned. In New Zealand social costs were higher than they needed to be, with resulting hardship and resentment.
- The results of decentralization and privatization demonstrate that the private sector can manage industrial forests more efficiently than the state. The investment in wood-processing facilities and improvements in business capability dramatically increased the profitability of the plantation forests.
- The New Zealand forestry sector, which exports most of its produce, is very susceptible to changes in the strength of the local currency against those of wood-importing nations and increased costs of production. Since 2003, these factors have overridden the impacts of the earlier reforms on forestry sector profitability.

There was no mention of other parts of the Pacific at the Davao meeting. In general, the tone of this workshop was not sanguine, since genuine devolution and participation of local communities were seen to be very rare.

The Interlaken meeting, held in Switzerland in 2004 and organized by the Swiss and Indonesian governments in support of the United Nations Forum on

Forests and co-hosted by CIFOR, was more global in its ambit: country cases included the US, Switzerland and Russia as well as developing states in Africa, Latin America and Asia. Its aims were also more specific, seeking to identify positive strategies to make the decentralization process work by examining the experience of countries where it was already advanced. There was an initial focus on federal systems; however, the type of decentralization involved there was revealed as not widely applicable (Gregersen et al, 2005). Ferguson and Chandrasekharan (2005) provided a further classification of approaches to decentralization and devolution, using several Asia-Pacific countries as examples: devolution to district governance (the Philippines and Indonesia); devolution to village governance (Korea and Nepal); decentralization involving customary ownership (New Guinea and Fiji); decentralization through privatization (China, Vietnam and Australia); and reversal of devolution in a federal system (again Australia). Melanesia, where the prevailing customary ownership systems differ from those in many other countries, was now represented.

Ribot took a more theoretical stance, focusing on institutions, representation, power and accountability. He listed practical questions to be asked by policymakers, activists and others about the kinds of institutions involved and the powers being transferred 'to provide the equity, efficiency, development and environment benefits that decentralization promises' (Ribot 2005, p100). Larson found problems especially in legal frameworks, in checks and balances on levels of power and authority, and in enforcement of regulations. Her overview of the inadequacies and inefficiencies of decentralization efforts led her to conclude that 'democratic decentralization is hard to come by' (Larson 2005, p55). Summing up, Capistrano and Colfer (2005) emphasized the gap between theory and practice. They noted the 'critical importance of the dynamic balance between authority, accountability mechanisms, responsibilities and revenue sharing across different levels of government' (Capistrano and Colfer, 2005, p301).

Although the conclusions from this workshop were similarly pessimistic, the emphases on power and accountability and the scale of management provided new insights. As noted in the foreword to the resulting book, 'The challenge is to find a governance framework that can balance the various local, national and global interests related to forests' (Wardojo and Roch, 2005). Such a challenge continued to confront us in the Yogyakarta meeting.

Dupar and Badenoch's approach

Dupar and Badenoch (2002), examining decentralization in the uplands of mainland Southeast Asia (Vietnam, Laos, Cambodia, Thailand and southern China), used a framework developed by Agrawal and Ribot (1999) to identify actors, powers and accountability relations and classify the type of decentralization at each site.[7] They also noted the prevalence of large, donor-assisted participatory projects across the region, especially in Laos, Vietnam and Cambodia. The fact that China led the way from 1980 made it easier for Vietnam and Laos to follow a decade or more later, although Dupar and Badenoch described the process in both Vietnam and Laos as 'deconcentration', with power relations remaining strictly hierarchical through the different tiers of government and each unit accountable to the one above.[8] China's household responsibility system, in which land was provided to households on long leases, was instead a form of privatization, though with strict controls on removal of products.

In their conclusion, Dupar and Badenoch stressed the importance of secure tenure, at either individual or community level, with strong mechanisms for downward accountability. Close social ties between leaders and the community were an important positive factor, enhancing social capital and likely to improve livelihoods, though livelihood and environmental objectives did not always converge. The development of independent mechanisms for conflict resolution could improve the likelihood of success in promoting both environmental sustainability and livelihood security.

Tenure: A Critical Factor

Who Owns the World's Forests? is the catchy title of a study by White and Martin (2002a) that attempts to assemble official tenure data for most of the world's top forested countries.[9] The authors distinguish between community lands, which are legally private property, from those considered to belong to the state, although in the latter case there is often a vast difference between official ownership and effective authority. In Africa and Asia 'governments own most forest land, but often appear to have authority over very little of it' (White and Martin, 2002a, p4).

Although only seven countries from Asia and the Pacific are represented, the variation among them is striking. It is clear that government landownership is by far the most dominant scenario, but this varies from around 100 per cent (Myanmar and Indonesia) to 3 per cent (Papua New Guinea). India stands out in the amount of public land reserved for community and indigenous groups, but a considerable area of private land is also in the hands of such groups in China, Australia and Papua New Guinea. The national governments usually prefer to lease forested land to large firms in concessions; that used to be true in the Philippines and is still the case in parts of Indonesia, Malaysia, Cambodia and Papua New Guinea. The current state of the forests as a result of concession activities is one reason for the shift to community forestry. Although the Philippines do not feature in White and Martin's figures, the legal recognition in 1997 of ancestral domain rights over more than a third of the former public forest represents a considerable triumph for the indigenous tribal communities, despite administrative and other problems in implementation. Traditional ownership in Papua New Guinea, on the other hand, is not without its difficulties, as unscrupulous firms have taken advantage of people's limited negotiating experience and promised benefits are seldom realized (see Box 3.1). Their plight is similar to that of their counterparts in Indonesian Papua, who have been recognized as traditional owners and offered logging rights up to 10,000ha if they apply for a licence from the Forestry Department and set up a cooperative (called *kopermas*). They are approached by Malaysian logging companies, who pay them ridiculously small sums for the rights to establish the *kopermas* and log the forests (Alhamid, 2004; EIA/Telapak, 2005).

In their concluding section, White and Martin (2002a, p22) suggest that 'the recognition of indigenous rights and community ownership – and the broader rationalization of public forest tenure – present an historic opportunity for countries to dramatically improve the livelihoods of millions of forest inhabitants'. A companion report lists the elements of tenure security. Most important is effective community institutions (including clear-cut and enforced rules and resource boundaries) (White and Martin, 2002b, p5). Legal activism for community claims, accompanied by community mapping of resources, public education and lobbying, are important strategies. It is also vital to make the best use of 'critical turning points ... during a transition to democracy' (White and Martin, 2002b, p14).

Such a turning point occurred most remarkably in Indonesia in 1998, with the fall of the Suharto regime. Although the subsequent rapid decentralization and changes to the Forest Law have provided opportunities to forest communities, there has also been frustration with the protracted negotiations necessary to change the mindset of government agencies.

Community-based Tenure in Indonesia

In 1993, the Indonesian Tropical Institute (LATIN) established the Community Forestry Consortium, which worked to document indigenous management systems in several provinces. Together with the World Agroforestry Centre (ICRAF) and the French research group ORSTOM, they assembled data concerning the indigenous agroforests at Krui (Lampung), based on the collection of damar resin. These dipterocarp forests (*Shorea javanica*) had been planted by generations of local Pesisir people and were managed by the *marga*, or communal group (Michon et al, 2000). From 1991 the Krui area had been declared state forest land and had come under threat from oil palm companies and logging interests. Biodiversity assessments, silvicultural studies and participatory mapping were assembled as evidence of the Krui system's environmental and social benefits. The Minister of Forestry was persuaded to consider this area a special case, and in January 1998 he declared Krui to be a KDTI (*kawasan dengan tujuan istimewa*, meaning 'special use zone') of 29,000ha. For the first time, local residents were permitted to harvest both timber and non-timber forest products from a state forest, as well as being granted unconditional management rights through a community governing structure. The ICRAF team drew on Philippine experience with certification of ancestral domain to serve as a blueprint for this measure.

It was hoped that this legislation would persuade the Ministry of Forestry to recognize the benefits from community-managed forests and move to a broader implementation of such policies, especially when, a few months later, the Suharto regime fell and citizens expected wide-ranging democratization. However, in this new period of 'reformation' the KDTI was perceived as too limited: it would have left the farmers dependent on the Ministry of Forestry. A follow-up study found that most farmers had never heard of the legislation yet felt reasonably secure in their tenure. The KDTI was neither implemented nor enforced, although it is now considered 'the most community-friendly tenure so far offered by the Ministry of Forestry' (Colchester et al, 2005, p19). Krui continues to be singled out as the one example of successful community forestry in Indonesia (Casson, 2005).

Following the fall of Suharto, the new Indonesian government rapidly pushed through decentralization laws (Numbers 22 and 25 of 1999) and a revised forestry law (Number 41 of 1999). A 1998 decree granting local cooperatives 35-year use rights in production, protection and conservation forests was changed in 2001, allowing the establishment of leases and permits over community forest (*hutan kemasyarakatan*, or HKm), but only for a maximum of 25 years[10] and outside nature reserves. One area where a number of HKm permits have been granted, largely as a result of campaigning by ICRAF, is in the coffee lands of Sumberjaya, also in Lampung.[11] Scientific studies have shown that multistrata coffee, grown in the shade under other planted trees, is reasonably protective of the hilly environment and the watershed of the Way Besai River (Verbist et al, 2004). Farmers must agree to guard remaining natural forest and adopt conservation techniques to minimize erosion. The farmers of Sumberjaya consider themselves better off once the system is in operation; the level of forest fires also decreases markedly: 'It is perceived land security, not actual title (or permit) which encourages farmer confidence and investment' (Colchester et al, 2005, p26; Suyanto et al, 2005, p1).

Another example of the slow evolution of community management has been in the recognition of indigenous rights in the large Kayan Mentarang National Park in East Kalimantan. This area of 1.4 million hectares is home to 16,000 Dayak people, whose lives and management of natural resources are regulated by *adat*, or customary law. In 1980 the area was originally proclaimed a nature reserve, in which no human activities were permitted. The Worldwide Fund for Nature (WWF) and the Indonesian Institute of Sciences (LIPI) carried out studies (including community mapping) from 1991 to 1997 to demonstrate community claims to the land and the forests. On the basis of those studies, the area became a national park, in which community activities are allowed. In their continuing struggle for recognition, the communities formed an inter-*adat* institution, the Alliance of the Indigenous Peoples of Kayan Mentarang National Park (FoMMA). With the WWF they lobbied the government to create a new model of collaborative national park management, and in this they achieved a compromise. A policy board, set up in 2002 with representatives from the central, provincial and district governments and the local communities, makes recommendations to the Minister of Forestry, though it does not have full management authority. Its challenge is to secure the economic future of the communities through rewards paid for environmental services (Eghenter et al, 2003).

The new forestry law permitted district governments to issue small-scale (100ha) timber concessions to firms that negotiated harvesting agreements with communities. This provided space for communities in provinces such as East Kalimantan to develop bargaining skills and 'self-enforce' their rights to secure profits from the forests (Engel and Palmer, 2006; Palmer and Engel, 2006). As Palmer and Engel demonstrate in a large study covering 65 communities, these activities were empowering: 'While the Ministry of Forestry still has formal ownership and land title to all of Indonesia's forests, this finding suggests that people already feel that they own the forest regardless' (Palmer and Engel, 2006, p28). The concessions allowed villagers to benefit for the first time from the proceeds of logging, though at times they also generated social conflict, both within and between communities, and caused damage to the forests. Other researchers have been less sanguine about the cash benefits obtained from the small concessions, at least in Malinau District (see Chapter 10). The benefits are described as being 'short-term, less than their potential and captured by the elite' (Wollenberg et al, 2006, p8), and the forests as degraded and fragmented. As many of these small concessions fell within larger concession boundaries, they were banned by the central forestry department in 2002, though they were still in operation in some districts until 2004.

The small gains through community forestry and the KDTI encouraged activists to push for further change. ICRAF researchers engaged in detailed analyses of Indonesia's 'forest zone' found that 25 per cent of it was used for agriculture or remained as grassland (Contreras-Hermosilla and Fay, 2005). They recommended that it be declassified from the 'forest zone' and titled. Such a process is seen as a critical first step toward recognition of indigenous and other community tenure.

Leasing forested lands to communities (like the existing community forests) would shield them from other claimants but provide rules for forest management. Eventually some leases could be converted to community landownership, using the Philippine model. The paper ends with a statement of national responsibility: 'While international assistance could be useful, in the end, progress will depend upon Indonesian leadership and political will' (Contreras-Hermosilla and Fay, 2005, p38).

Community-based Tenure in Other Countries

Malaysia presents a different scenario in legal terms, though the results are similar. The category 'native customary land' (NCL), instituted in Peninsular Malaysia in colonial times to protect indigenous Malay land from the rubber industry, was later extended to the Borneo states of Sabah and Sarawak, where those who could prove their indigenous status could register their land. Much of the land so designated was still forested, and on it local people could continue traditional activities. However, under modern pressures for 'development', these lands have increasingly been sought for alternative uses by outsiders, for both logging and plantation crops. In Sabah, plantation and other interests may secure country leases and temporary occupation licences for unregistered NCL. Even when people apply to register their lands, applications may take decades to process, especially where claims are overlapping. Frustrated locals sometimes invite logging companies to secure a temporary occupation licence and remove timber in exchange for part of the proceeds. Vaz (2006), involved in a partnership between a community and a conservation NGO in a remote upland area of Sabah, noted that it was only through the rapid creation of a communal 'native reserve' that a large area of excellent forest could be secured from the imminent threat of logging and its resources safeguarded for domestic access and possible tourism development. Negotiations were difficult because powerful interests within the village were more focused on obtaining profits from logging. The villagers were able to retain their communal forests with the collaboration of sympathetic government officers as well as WWF Malaysia, though the need to improve livelihoods remained unresolved.

Building on village traditions that had survived the disturbances of the previous 20 years, community forestry in Cambodia was started by NGOs in the early 1990s and grew rapidly. It is, however, still largely dependent on foreign donors and NGOs, and the degraded forests generate little income for participants (Sunderlin, 2006). The Cambodian government in 2004 promulgated a decree on community forest management that should have provided some security of tenure. Many villagers had previously lost land to logging concessions in the flurry of 'anarchic logging' during the 1990s. Analogous to the Indonesian situation, their rights are currently threatened by expansion of planted tree crops, in this case large-scale rubber plantations (NGO Forum, 2004).

One Southeast Asian state that still seems to be going against the trend towards establishing community forests is Thailand, even though the past 14 years have seen various drafts of a community forests bill. Upland dwellers found their land declared 'conservation forest' as a result of the Royal Forest Department's policy to place 25 per cent of the country under conservation.[12] In northern Thailand the conservation zone covers 48 per cent of the area, reaching 80 per cent in some provinces (Zurcher, 2005). The livelihood insecurity of upland dwellers, many of whom belong to ethnic minorities, is compounded by persistent threats of relocation to lowland areas. Activist groups have taken up the farmers' cause for legal recognition of community forest rights, mounting vigorous arguments to counter urban environmentalists, who are largely blamed for the failure of the community forests bill in the Thai Parliament.[13] Walker (2004) has drawn attention to the widespread commercialization of upland agriculture, which does not match the image of the farmer as a forest dweller leading a largely subsistence lifestyle. What he terms the 'arborealization' of agriculture by activist groups has led to the stereotype of farmer-forager practising shifting cultivation and collecting forest products. The bill would not permit farming in community forests, so the farmers could be worse off. A better strategy would be to acknowledge the role of the market in upland farming and to remove commercially farmed lands from the conserved forest. As with the Indonesia case, revision of land management categories is urgently required.

Tenure and forest fire

Both fire incidence and its suppression are often linked to the tenure status of local communities. For example, in the series of eight studies conducted across Sumatra and Kalimantan (Indonesia) in 1999 and 2000 by a joint team from CIFOR and ICRAF, by far the most significant cause of fire was burning related to land tenure and land-use conflict and competition. People would burn to establish property claims or drive away competing occupants (Dennis et al, 2005). As already noted, improved tenure status in Sumberjaya (Sumatra) reduced fire incidence (Suyanto et al, 2005). Fire suppression is also affected by tenure: if people perceive that the land belongs to others (including the government), they have little incentive to put fires out. Although traditional methods of fire control were once widely practised, for example in northern Thailand and northeastern India, the intrusion of government authority eroded such systems, as people felt the responsibility was no longer theirs (Darlong, 2002; Pearmsak Makarabhirom et al, 2002). Allocating forests to communities has been recommended as the most effective means to control fire in Vietnam (Ha Thi Linh, 2002).

Tenure and illegal logging

In Indonesia, processing capacity and industrial demand for wood products far outweigh the supplies of timber from legal sources, such as concessions, which means that much internal trade is technically illegal. This is especially true in provinces such as Riau, where large pulp and paper companies continue to obtain fibre mainly from natural forests (Potter and Badcock, 2001; Barr, 2001). Illegal logging, though not a new phenomenon, has increased greatly over the past decade and has many manifestations (Casson and Obidzinski, 2002; FAO, 2005b; Obidzinski, 2005). Institutional attempts to control it are being put in place at all levels, from local efforts to exclude outsiders from community forests, to government policing, aided by international organizations such as Forest Law Enforcement and Governance, and global private initiatives to certify 'legally' sourced timbers. With many opportunities for large profits still available to government officers, together with smaller amounts for local, national and regional entrepreneurs, this question is complex. One must examine the political economy of illegal forestry activities at different scales. As Dudley (2002, p168) points out, a villager's view of illegal logging is affected by community dissatisfaction with government forest policy and a perception that access to forest resources is being threatened. Many villagers, whose rights in their traditional forests are not recognized, depend on 'illegal' logging to survive and may be adversely affected by forest law enforcement (Colchester et al, 2006).

In Vietnam the state largely blames local people's illegal logging for its recent high deforestation rates. Vietnam has a partial logging ban, with small quotas being set each year for legal felling; the high demand for timber and wood products, as in Indonesia, is met by 'illegal' logging. The state had control over most forest areas and, until 2004, would not allow stocked forests to be managed by communities. After the land law was revised in 1993, small areas of mostly bare land were given to households (20-year leases for farmlands and 50-year forest leases). The law was changed in October 2004 to allow for allocation of forest land to communities (Tan Quang Nguyen, Chapter 12 of this volume), though much larger areas of forest still remained under state forest enterprises. These have been logged intensively, rapidly reducing the standing volume of timber. The enormous profits from the wood trade have led to corruption at all levels. Since many households still have no forest land, they justify their activities through a sense of entitlement and social justice (McElwee, 2004, p112). As in Indonesia, the problem is related to questions of tenure and poverty as well as trade.

Equity Issues: Gender and Caste

In the literature on community forest management, there is generally some mention of equity, but few studies directly address the important questions of gender, caste and class. The most important research on this topic comes from South Asia, especially India and Nepal.

Decentralization, though likely to expand the scope of citizens' participation in governance, will not by itself ensure that women and men are represented on an equal basis. Where decentralization confers power on 'traditional' authorities, such as tribal elders, it may be inimical to women's interests (Molyneux and Razavi, 2005). Referring to Indonesia, Siahaan (2002) argues that decentralization does not work in a gender-neutral framework: local governments face structural and cultural obstacles to realizing gender equity. The revival of local patriarchal values and institutional constraints pose great challenges to women, who have higher rates of illiteracy and unemployment than men. Increasing the cost-effectiveness of participatory development planning and improving women's self-confidence means involving women.

What Agarwal (2001) has termed 'participatory exclusions' in her studies on India and Nepal is equally a problem in mainland Southeast Asia (Dupar and Badenoch, 2002), the Philippines (Edmunds and Wollenberg, 2003) and Indonesia (Siahaan, 2002). Agarwal first presents a typology of participation, from nominal group membership through passive participation to active and finally empowering participation that influences a group's decisions. As she then argues, most women are not even participating at the nominal level and have no membership in community forest groups. India's Joint Forest Management committees are far from democratic: women are either absent from the discussions or have no influence on decisions, despite being responsible for everyday forest protection activities and highly dependent on collection of forest resources, such as firewood, grass and certain foods. When a forest is closed to allow regeneration, women may be forced to travel long distances to find new sources. Sarin et al (1998) present case studies of the severe impact of forest closures on very poor women, especially fuelwood head-loaders. Moreover, women often know more than men about particular tree species and other aspects of the local environment. Women's exclusion from decision-making has implications for equity, efficiency and forest planning, and to gain entry, women must learn to bargain with the state, the community and the family (Agarwal, 2001). Under pressure from external agents, the Indian state has more recently introduced rules for minimal female participation (at least 50 per cent representation of women in the Joint Forest Management general body and 33 per cent in the executive).

Though echoing Agarwal's findings in her own research in Orissa, Arora-Jonsson concluded that women preferred to work through their own groups when it came to local development and resource management, since they had more space to exercise agency in those contexts. Women's savings and credit groups enabled them to carry out collective activities, such as planting trees, tending backyard plantations or cleaning village commons. Given the deeply entrenched nature of patriarchy in India, she questioned whether women in fact wanted to join formal associations or committees (Arora-Jonsson, 2005, p28, and Chapter 4 of this volume). Women often took on activist roles, supporting men's organizations with marches and blockades, at times challenging both the Forestry Department and illegal loggers. However, talk of a wider network, in which issues such as dowry payments and domestic violence might be raised, made the men uneasy. Plans for a federation of women's groups were eventually stopped by withdrawal of funding: 'the dreams of a federation are still distant, perhaps because it is still too threatening' (Arora-Jonnson, 2005, p105).

A study by Nightingale (2005) provides a nuanced analysis of the interrelationships between gender, caste, work and the forest in a forest user group in northwestern Nepal. Nightingale focuses on caste and the differentiation of knowledge, with the district forest office being the repository of 'scientific forest management', which is regarded as superior and transmitted to the village through the high caste, literate male elite. The adoption of such practices is used both to gain favour with the district office and to assert authority within the group. However, such forest practices potentially undermine three objectives of community forestry: a pro-poor orientation, promotion of democratic institutions and maintenance of ecosystem health (Nightingale, 2005, p584). Although the literate, high-caste men claimed the right to control the forest user group and manage the forest, it was the illiterate, low-caste men and illiterate women of all castes who did most of the forest work. Women were able to challenge the elite through their practical understanding of the forest. They prevented the men from trying to introduce permits for collection of firewood and leaf litter, resources harvested almost daily. Although they had important contributions to make, however, they subordinated their knowledge to that of the elite. The management techniques promoted were for production of timber, not the multi-species 'subsistence' harvesting needed by the village. As noted in the discussion on India, scientific forestry cannot produce resources needed for daily survival, yet it is being promoted in community forests (see Chapter 9).

In the examples from both Arora-Jonsson and Nightingale, marginalized groups are still able to find ways of asserting their own agency and fighting back against the restrictions of a social system that is slow to change. Dupar and Badenoch (2002) suggested that targeted development interventions combining requirements for participation of marginalized groups with informal education to build confidence were likely to be most successful. In Nepal, some women have taken over management responsibility of forest user groups from men: Poffenberger (2000) indicated that 30 per cent were run by women, including some of the most active.[14] A recent overview suggests that 'equity and gender issues are now being emphasized in donor-funded projects and the training of [district forest office] staff' (Katila, 2005).

Conclusions

Hierarchical, command-and-control approaches to forest governance, in which central governments developed policy and their forestry departments implemented it (called 'old governance'), have given way to more democratic arrangements emphasizing collaboration and interdependence between the state and civil society ('new governance'). New governance usually includes some measure of decentralization and devolution (Gluck et al, 2005). It also needs to coordinate the involvement of a variety of private and public actors with different objectives – the local community, non-governmental organizations (both local and international), international research institutes, private firms, traders and others in national and international marketing chains, and members of local, regional and national government agencies. Such groups constitute a 'policy network' with a great diversity of values, but flexibility and trust are of high importance, the aim being 'to steer multiple actors towards public goals such as sustainable forest management' (Gluck et al, 2005, p54). Moving across scale from community to local and central government and back and identifying appropriate roles for each level become easier when interested intermediaries are in place to build capacity and smooth the path.

To what extent do these ideas fit the reality of the situation in the Asia-Pacific forests? It is worth remembering that before the old governance regimes developed, with the spread of colonialism and the concepts of scientific forestry, earlier traditional forms of governance held sway, under which local communities managed their forests according to their own rules. Both forms are still evident in parts of the region today, and traditional forms are still powerful in areas remote from central government influence. Decentralization can have the effect of destabilizing such systems but not offering coherent alternatives, sometimes replicating remnants of the old command-and-control systems (still quite prominent in India, Nepal and Myanmar), at other times producing only anarchy and corruption by local elites. Cambodia, Indonesia and Vietnam have certainly experienced the latter outcome, which appears more prominent in states with high demand for forest products and easy opportunities for illegal logging and smuggling beyond national borders.

The role of intermediaries in policy networks is highly important, especially in the early period of decentralization and devolution. NGOs and international research groups have been the catalyst in securing more community-friendly tenure arrangements in Indonesia, simultaneously reducing the incidence of forest fire. Though such organizations have their own agendas, they are nonetheless needed until local institutions are strong enough to negotiate with all stakeholders on an equal basis. The urgent need for strong and secure tenure is especially vital where corporate interests can override traditional arrangements.

Equity issues within local communities are often overlooked in the establishment of local institutions, which may quickly be dominated by male elites. As detailed studies indicate, women may initially have to seek 'underground' ways of influencing decisions, though legislation mandating representative membership would at least ensure adequate flows of information to those most affected.

Although the kinds of decentralization and devolution identified in this overview remain inadequate, messy, slow in implementation and often disappointing to their participants, nevertheless there are signs of improvement in some areas. The impetus and the zeal certainly exist for better systems of forest management with real devolution of power to forest communities. The people will be satisfied with nothing less.

Notes

1 The countries under study include those in the FAO categories of East Asia, South and Southeast Asia and Oceania. West and Central Asia, which collectively have a small forest area, are omitted.

2 Using figures from *Bisnis Indonesia* of illegal timber exports from Papua at 600,000m^3/month in 2002, Alhamid calculated that the income from that source of 7.2 trillion rupiah per year was five times greater than the provincial budget and three times the budget of the Indonesian military (Alhamid, 2004, p273).

3 From an extension of traditional *tumpangsari* systems in Java (where food crops are planted among regenerating teak trees for the first two years), to village woodlots planted in India to take pressure off natural forests and family-based agroforest homelots in the Philippines. In Thailand the forest village programme was adopted in 1978, the same year as Nepal's Forest Act was amended to permit the handover of degraded forest land to local *Panchayats*.
4 That 'narrative' of deforestation ignored villagers' traditional contribution to forest maintenance (Arora-Jonssen, 2005).
5 Emphasis in original.
6 Emphasis added.
7 Fujita and Phengsopha's study on Laos (Chapter 8 of this volume) uses a similar framework.
8 Under the definition provided by Fisher et al, this process would also be 'decentralization without devolution'.
9 Data are available for 24 of the top 30 forested countries, representing 93 per cent of the world's natural forests.
10 Initially, farmers receive a six-year provisional permit.
11 It is interesting that some of these permits are in protected forest, from which the Indonesian government had previously evicted coffee growers.
12 In 1985 the target of permanent forest cover was set at 40 per cent, of which 15 per cent was conservation forest and 25 per cent production forest. After floods in the south in 1989, which led to a logging ban, those figures were reversed. In 1992 the agency subdivided the forest reserves for agricultural, economic and conservation purposes (Zurcher, 2005).
13 The last rejection in the Senate was in 2002.
14 The Nepal–Australia Forestry Project, which operated in two districts over 25 years, had shifted its focus to expanding women's literacy in its final years of operation in order to support women's involvement in the forest user groups.

REFERENCES

Adhikari, B., Williams, F. and Lovett, J. (2006) 'Local benefits from community forests in the middle hills of Nepal', *Forest Policy and Economics*, vol 9, no 5, pp464–478

Agarwal, B. (2001) 'Participatory exclusions, community forestry and gender: An analysis for South Asia and a conceptual framework', *World Development*, vol 29, no 10, pp1623–1648

Agrawal, A. and Ribot, J. (1999) 'Accountability in decentralization: A framework with South Asian and West African environmental cases', *The Journal of Developing Areas*, vol 33, no 4, pp473–502

Alhamid, H. (2004) 'Forests for the people? Indigenous forest management under decentralization: A case study of the Rendani Protection Forest, Papua, Indonesia', unpublished PhD thesis, School of Resources, Environment and Society, Australian National University, Canberra

Arora-Jonsson, S. (2005) 'Unsettling the order: Gendered subjects and grassroots activism in forest communities', unpublished PhD thesis, Swedish University of Agricultural Sciences, Uppsala, Sweden

Banerjee, A. (2000) 'Devolving forest management in Asia-Pacific countries', in T. Enters, M. Victor and P. Durst (eds) *Decentralization and Devolution of Forest Management in Asia and the Pacific*, RECOFTC Report No 18 and RAP Publication 2000/1, The Regional Community Forestry Training Center for Asia and the Pacific (RECOFTC), Bangkok

Barr, C. (2001) *Banking on Sustainability: Structural Adjustment and Forestry Reform in Post-Suharto Indonesia*, WWF Macroeconomics Program Office and CIFOR, Washington, DC

Bun, Y., King, T. and Shearman, P. (2004) *China's Impact on New Guinea's Forestry Industry*, Forest Trends, Washington, DC

Capistrano, D. and Colfer, C. J. P. (2005) 'Decentralization: Issues, lessons and reflections', in C. J. P. Colfer and D. Capistrano (eds) *The Politics of Decentralization: Forests, Power and People*, Earthscan, London

Casson, A. (2005) 'Cat's eye forests: The Krui damar gardens', in P. Durst, C. Brown, H. Tacio and M. Ishikawa (eds) *In Search of Excellence: Exemplary Forest Management in Asia and the Pacific*, FAO and RECOFTC, Bangkok

Casson, A. and Obidzinski, K. (2002) 'From new order to regional autonomy: Shifting dynamics of "illegal logging" in Kalimantan, Indonesia', *World Development*, vol 30, no 12, pp2133–2151

Clarke, M. (2000) 'Devolving forest ownership in New Zealand: Processes, issues and outcomes', in T. Enters, M. Victor and P. Durst (eds) *Decentralization and Devolution of Forest Management in Asia and the Pacific*, RECOFTC Report No 18 and RAP Publication 2000/1, RECOFTC, Bangkok

Colchester, M., Eka Dinata, A., Fay, C. C., Pasya, G., Situmorang, L., Sirait, M. T., van Noordwijk, M., Cahyaningsih, N., Budidarsono, S., Suyanto, S., Kusters, K. Manalu, P. and Gaveau, D. (2005) 'Facilitating agroforestry development through land and tree tenure reforms in Indonesia', ICRAF Southeast Asia Working Paper No 2005-2, ICRAF, Bogor, Indonesia

Colchester, M., Boscolo, M., Contreras-Hermosilla, A., Gatto, F. D., Dempsey, J., Lescuyer, G., Obidzinski, K., Pommier, D., Richards, M., Sembiring, S. S., Tacconi, L., Rios, M. T. S. and Wells, A. (2006) 'Justice in the forest: Rural livelihoods and forest law enforcement', *Forest Perspectives*, vol 3, CIFOR, Bogor, Indonesia

Contreras-Hermosilla, A. and Fay, C. (2005) *Strengthening Forest Management in Indonesia through Land Tenure Reform: Issues and Framework for Action*, Forest Trends, Washington, DC

Darlong, V. (2002) 'Traditional community based fire management among the Mizo shifting cultivators of Mizoram in northeast India', in P. Moore, D. Ganz, L. C. Tan, T. Enters and P. Durst (eds) *Communities in Flames: Proceedings of an International Conference on Community Involvement in Fire Management*, FAO, Bangkok

Dennis, R. A., Mayer, J., Applegate, G. B., Chokkalingam, U., Colfer, C. J. P., Kurniawan, I., Lachowski, H., Maus, P., Permana, R. P., Ruchiat, Y., Stolle, F., Suyanto, S. and Tomich, T. (2005) 'Fire, people and pixels: Linking social science and remote sensing to understand underlying causes and impacts of fire in Indonesia', *Human Ecology*, vol 33, no 4, pp465–504

Dudley, R. (2002) 'Dynamics of illegal logging in Indonesia', in C. J. P. Colfer and I. A. P. Resosudarmo (eds) *Which Way Forward? People, Forests and Policymaking in Indonesia*, Resources for the Future/CIFOR, Washington, DC

Dupar, M. and Badenoch, N. (2002) *Environment, Livelihoods and Local Institutions: Decentralization in Mainland Southeast Asia*, World Resources Institute, Washington, DC

Edmunds, D. and Wollenberg, E. (eds) (2003) *Local Forest Management*, Earthscan, London

Eghenter, C., Labo, M. and Ferrari, M. F. (2003) 'Indonesia: The Dayak people in the first co-managed protected area', *Community Forestry E-news*, 31 August 2003, No 2003:14, RECOFTC, Bangkok

EIA/Telapak (2005) *The Last Frontier: Illegal Logging in Papua and China's Massive Timber Theft*, EIA/Telapak, London, Washington, DC, and Bogor, Indonesia

Engel, S. and Palmer, C. (2006) 'Who owns the right? The determinants of community benefits from logging in Indonesia', *Forest Policy and Economics*, vol 8, pp434–446

FAO (2005a) *Global Forest Resources Assessment 2005*, FAO, Rome

FAO (2005b) 'Best practices for improving law compliance in the forestry sector', FAO Forestry Paper No 145, FAO, Rome

Ferguson, I. and Chandrasekharan, C. (2005) 'Paths and pitfalls of decentralization for sustainable forest management: Experiences of the Asia Pacific region', in C. J. P. Colfer and D. Capistrano (eds) *The Politics of Decentralization: Forests, Power and People*, Earthscan, London

Fisher, R. (2000) 'Decentralization and devolution in forest management: A conceptual overview', in T. Enters, M. Victor and P. Durst (eds) *Decentralization and Devolution of Forest Management in Asia and the Pacific*, RECOFTC Report No 18 and RAP Publication 2000/1, RECOFTC, Bangkok

Gautam, A., Shivakoti, G. and Webb, E. (2004) 'A review of forest policies, institutions, and changes in the resource condition in Nepal', *International Forestry Review*, vol 6, no 2, pp136–148

Gibbs, C., Payuan, E. and del Castillo, R. (1990) 'The growth of the Philippines social forestry program', in M. Poffenberger (ed) *Keepers of the Forest: Land Management Alternatives in Southeast Asia*, Kumarian Press, New York

Gilmour, D. (2003) 'Retrospective and prospective view of community forestry in Nepal', *Journal of Forest and Livelihood*, vol 2, no 2, pp5–7

Global Witness (2003) 'A conflict of interests: The uncertain future of Burma's forests', Global Witness briefing document, Global Witness, London

Global Witness (2005) 'A choice for China: Ending the destruction of Burma's northern frontier forests', Global Witness briefing document, Global Witness, London

Gluck, P., Rayner, J. and Cashore, B. (2005) 'Changes in the governance of forest resources', in G. Mery, R. Alfaro, M. Kanninen and M. Lobovokov (eds) *Forests in the Global Balance – Changing Paradigms*, International Union of Forestry Research Organizations (IUFRO) World Series, vol 17, IUFRO, Vienna, pp51–74

Gregersen, H., Contreras-Hermosilla, A., White, A. and Phillips, L. (2005) 'Forest governance in federal systems: An overview and implications for decentralization', in C. J. P. Colfer and D. Capistrano (eds) *The Politics of Decentralization: Forests, Power and People*, Earthscan, London

Ha Thi Linh (2002) 'Experiences in community forest fire control in Viet Nam', in P. Moore, D. Ganz, L. C. Tan, T. Enters and P. Durst (eds) *Communities in Flames: Proceedings of an International Conference on Community Involvement in Fire Management*, FAO, Bangkok

Katila, P. (2005) 'Experiences from community forestry in Nepal', in G. Mery, R. Alfaro, M. Kanninen and M. Lobovokov (eds) *Forests in the Global Balance – Changing Paradigms*, International Union of Forestry Research Organizations (IUFRO) World Series, vol 17, IUFRO, Vienna, p218

Larson, A. (2005) 'Democratic decentralization in the forestry sector: Lessons learned from Africa, Asia and Latin America', in C. J. P. Colfer and D. Capistrano (eds) *The Politics of Decentralization: Forests, Power and People*, Earthscan, London

McElwee, P. (2004) 'You say illegal, I say legal: The relationship between "illegal" logging and land tenure, poverty and forest use rights in Vietnam', *Journal of Sustainable Forestry*, vol 19, nos 1–3, pp97–135

Menzies, N. (2002) 'Global gleanings: Lessons from six studies of community-based forest management', report prepared for the Ford Foundation's Environment and Development Affinity Group (EDAG)

Menzies, N. (2004) 'Communities and their partners: Governance and community-based forest management', *Conservation and Society*, vol 2, no 2, pp449–456

Michon, G., de Foresta, H., Kusworo, A. and Levang, P. (2000) 'The damar agroforests of Krui, Indonesia: Justice for forest farmers', in C. Zerner (ed) *People, Plants and Justice: The Politics of Nature Conservation*, Columbia University Press, New York

Milne, G. (2005) *Unlocking Opportunities for Forest-Dependent People in India* (2 vols), World Bank, Washington, DC

Molyneux, M. and Razavi, S. (2005) 'Beijing plus ten: An ambivalent record on gender justice', *Development and Change*, vol 36, no 6, pp983–1010

NGO Forum (2004) Cambodia [Forest Livelihoods Project], available at www.ngoforum.org.kh/Land/Livelihood/background.htm, accessed 4 April 2006

Nightingale, A. (2005) '"The experts taught us all we know": Professionalisation and knowledge in Nepalese community forestry', *Antipode*, vol 37, pp581–604

Obidzinski, K. (2005) 'Illegal logging in Indonesia: Myth or reality?', in B. Resosudarmo (ed) *The Politics and Economics of Indonesia's Natural Resources*, Institute of Southeast Asian Studies (ISEAS), Singapore

Palmer, C. and Engel, S. (2006) 'For better or for worse? Local impacts of the decentralization of Indonesia's forest sector', working paper, Eidgenosische Technische Hochschule (ETH), Zurich, Switzerland

Pearmsak Makarabhirom, D. and Surin Onprom (2002) 'Community involvement in fire management: Cases and recommendations for community-based fire management', in P. Moore, D. Ganz, L. C. Tan, T. Enters and P. Durst (eds) *Communities in Flames: Proceedings of an International Conference on Community Involvement in Fire Management*, FAO, Bangkok

Poffenberger, M. (ed) (1990) *Keepers of the Forest: Land Management Alternatives in Southeast Asia*, Kumarian Press, New York

Poffenberger, M. (ed) (2000) *Communities and Forest Management in South Asia*, IUCN (Working Group on Community Involvement in Forest Management), Gland, Switzerland

Potter, L. and Badcock, S. (2001) *The Effect of Indonesia's Decentralisation on Forests and Estate Crops in Riau Province: Case Studies of the Original Districts of Kampar and Indragiri Hulu*, Case Studies 6 and 7 in 'Case studies on decentralization and forests in Indonesia' series, CIFOR, Bogor, Indonesia

Pragtong, K. and Thomas, D. (1990) 'Evolving management systems in Thailand', in M. Poffenberger (ed) *Keepers of the Forest: Land Management Alternatives in Southeast Asia*, Kumarian Press, New York

Ribot, J. (2005) 'Choosing representation: Institutions and powers for decentralized forest management', in C. J. P. Colfer and D. Capistrano (eds) *The Politics of Decentralization: Forests, Power and People*, Earthscan, London

Sarin, M., Raju, M. S., Chatterjee, M., Banarjee, N. K. and Hiremath, S. (1998) *Who is Gaining? Who is Losing? Gender and Equity Concerns in Joint Forest Management*, Society for the Promotion of Wastelands Development, New Delhi

Siahaan, A. (2002) 'Women and decentralization in Indonesia: Bringing local government closer to women?', policy paper, available at www.policy.hu/siahaan/Policypaper1.htm, accessed 3 March 2006

Stoney, C. and Bratamihardja, M. (1990) 'Identifying appropriate agroforestry technologies in Java', in M. Poffenberger (ed) *Keepers of the Forest: Land Management Alternatives in Southeast Asia*, Kumarian Press, New York

Sunderlin, W. (2006) 'Poverty alleviation through community forestry in Cambodia, Laos and Vietnam: An assessment of the potential', *Forest Policy and Economics*, vol 8, no 4, pp386–396

Suyanto, S., Permana, R., Khususiyah, N. and Joshi, L. (2005) 'Land tenure, agroforestry adoption, and reduction of fire hazard in a forest zone: A case study from Lampung, Sumatra, Indonesia', *Agroforestry Systems*, vol 65, pp1–11

Turia, R. (2005) 'Cannot see the land for the trees: The forest management dilemma in Papua New Guinea', unpublished PhD thesis, Department of Human Geography, Australian National University, Canberra

Upreti, B. and Srestha, B. (2000) 'Balancing power in community forestry: Decentralization and devolution of power', in T. Enters, M. Victor and P. Durst (eds) *Decentralization and Devolution of Forest Management in Asia and the Pacific*, RECOFTC Report No 18 and RAP Publication 2000/1, RECOFTC, Bangkok

Vaz, J. (2006) 'Seeking space for biodiversity through improved tenure security for local communities: A community-conservation NGO partnership in Sabah', in F. Majid Cooke (ed) *State, Communities and Forests in Contemporary Borneo*, Asia-Pacific Environment Monograph I, Australian National University Press, Canberra

Verbist, B., Ekadinata Putra, A. and Budidarsono, S. (2004) 'Penyebab guna lahan dan akibatnya terhadap fungsi dearah aliran sungai (DAS) pada lansekap agroforestri berbasis kopi di Sumatera' ('Causes and effects of land use change on watershed functions in a coffee agroforestry landscape in Sumatra'), *Agrivita*, vol 26, pp29–38

Wardojo, W. and Roch, P. (2005) 'Foreword', in C. J. P. Colfer and D. Capistrano (eds) *The Politics of Decentralization: Forests, Power and People*, Earthscan, London

Walker, A. (2004) 'Seeing farmers for the trees: Community forestry and the arborealisation of agriculture in northern Thailand', *Asia Pacific Viewpoint*, vol 45, no 3, pp311–324

White, A. and Martin, A. (2002a) *Who Owns the World's Forests? Forest Tenure and Public Forests in Transition*, Forest Trends, Washington, DC

White, A. and Martin, A. (2002b) *Strategies for Strengthening Community Property Rights over Forests: Lessons and Opportunities for Practitioners*, Forest Trends, Washington, DC

White, A., Xiufang Sun, Canby, K., Jintao Xu, Barr, C., Katsigris, E., Bull, G., Cossalter, C. and Ilsson, S. (2006) *China and the Global Market for Forest Products: Transforming Trade to Benefit Forests and Livelihoods*, Forest Trends, Washington, DC

Wollenberg, E., Moeliono, M., Limberg, G., Iwan, R., Rhee, S. and Sudana, M (2006) 'Between state and society: Local governance of forests in Malinau, Indonesia', *Forest Policy and Economics*, vol 8, no 4, pp421–433

Zurcher, S. (2005) 'Public participation in community forest policy in Thailand: The influence of academics as brokers', *Geografisk Tidsskrift, Danish Journal of Geography*, vol 105, no 1, pp77–88

4

A Different Vantage Point: Decentralization, Women's Organizing and Local Forest Management

Seema Arora-Jonsson

The past decade has witnessed an important shift in international and national discussions about natural resource management (see, for example, Shackleton et al, 2002). The forest sector especially has undergone policy shifts wherein people's participation and the attention to gender relations in forest issues have been recognized as important in bringing about sustainable forest governance. Decentralization has been considered a powerful means by which to achieve development goals in ways that respond to the needs of local communities (cf. World Bank, 2000). Scholars, development practitioners and others working with resource management have advocated linking up with efforts by men and women in local communities to form viable local institutions for the governance of natural resources. Keeping in mind the importance of gender relations in everyday life, in this chapter I discuss *how* a more democratic and gender-equitable decentralization might be achieved in natural resource management. My aim is less to present firm solutions or recommendations than to examine some of the problems and contradictions that can arise in efforts toward increased local management. By suggesting ways to approach them, I hope to initiate a discussion that can inform theoretical and policy debates on gender-equitable natural resource management. More importantly, I hope that the discussion may encourage thinking and actions that each of us can use to benefit both men and women living and working with development and natural resources in their communities.

Using examples from research in villages in Orissa, India, and a village in northwestern Sweden, I invite readers to think about alternative ways of organizing for decentralization. The examples draw attention to the process and the 'blind spots' of decentralizing efforts – that is, to the social and gender relations that are so taken for granted that they become invisible and lead us to disregard the important ways in which they structure the process and determine outcomes. The two study sites, in such different parts of the world, were chosen in order to understand local efforts for resource management from a North–South perspective, in a context where discussions on resource management are becoming increasingly 'global' and where agencies from the North play a critical role in debates in the South.

I examine the different meanings given to local management and development and discuss what sustainable local forestry might mean in the everyday life of the people directly affected by efforts aimed at transferring decision-making power and responsibility for the forests. Both the Indian and Swedish examples are grass-roots efforts, not the result of decentralization initiated by government authorities. Yet the local communities' struggles for management rights offer important lessons for processes of decentralization. In both case study sites, village communities had formed organizations to manage forests locally and to work on development issues. Not finding a place for themselves in these organizations, women in both places chose to organize themselves in their own collectives. The analysis of the relationship between the women's organizing and the community forest associations as well as the role of outside agencies in supporting local management and women's efforts in these two different communities provide interesting parallels. These parallels generate insights that go beyond the two specific contexts to raise crucial questions about the gendered nature of natural resource management and decentralization.

The case studies were carried out from the vantage point of the women's groups, not from the perspective of the local forest management organizations, as is normally the case when studying local resource management. Such an approach brought to light three important aspects that I describe here as the blind spots of decentralization:

1 *Social nesting of forest governance.* The case studies illustrate the ways in which forest management is nested in and linked inextricably with other issues in everyday community life. This highlights the impossibility of addressing forest management without relating it to other community issues, since the relationships between these issues and the people involved play an important part in determining forest management outcomes. This is very much a gendered question, partly because of the sexual division of labour in most villages, but also because the issues and activities that get prioritized as 'community activities' depend on the gender and power relations within the village and on relationships with development practitioners and bureaucrats from outside.
2 *Unintended effects of inclusion.* Even with the best of intentions, attempts to include women in forest management organizations to ensure more effective and gender-equitable outcomes can have a detrimental effect if attention is not also directed to how different women choose to participate or choose to organize themselves in relation to these attempts. The assumption is often made that getting women to meetings takes care of women's participation. As the

two cases show, such an approach may end up undermining efforts towards participation and sustainable forest management, and new and varied means of collaboration must therefore be considered.

3 *The skewing effects of outsiders.* In a process of decentralization, development practitioners and other external agents may confront a multiplicity of groups that present themselves as the representatives of the 'community'. In the two case studies, I illustrate how the outsiders shifted the direction of the processes that started in the villages. Their relationships with the villagers and among themselves played as much a part in forest governance as did community efforts. As researchers, development agents or government officials, we play an extremely important role, but how we may inadvertently be reinforcing unequal relationships is rarely the subject of analysis.

In the rest of the chapter, I first briefly outline the research approach and fieldwork methods that brought into focus for me the blind spots of local management that would otherwise have remained invisible. Following the description of the methods, I turn to the broader context of local forest management – that is, the social relations of forest governance. I then look at the inclusion of women in forest organizations and finally at the role of outside actors in development projects. The chapter concludes with a discussion of the main points.

METHODOLOGY

I started my research with a conventional approach to research on local management initiatives – in other words by focusing on the formal institutions that work with local management and development. However, preliminary studies in both sites showed that women were absent from these organizations. Rather than focusing on why women were absent and how they might be included in forest organizations, I chose instead to study the spaces where women were present, how they themselves chose to work in relation to the forests and what they wanted from local control. By starting outside the mainstream organizations, I was able to question the norms and the assumptions on which local management initiatives were based, as well as to understand local management in its broader context. This approach brings to light a different knowledge about forest management, about how the norms of decentralization that appear to open up spaces for people may in fact also become the basis of discrimination. The focus and methodology thus highlight the limitations of the knowledge that we have and illustrate the need to open up to different approaches and think about how they can inform our praxis.

This analysis is based on studies carried out in Nayagarh District of Orissa, India, first in 1993 and then in 1998 and 1999. Although I focus on my research in Orissa, I relate it to studies carried out in a village in the sparsely populated region of northwestern Sweden between 1998 and 2003, a context in the 'developed' world where efforts for local management of forests are now being debated. The research was based on group interviews with women's groups; individual interviews with women, men, the staff of the women's development programme in Nayagarh, and men in the forest organizations and village association in both places; and participant observation at meetings and other get-togethers. The major part of the research in Sweden was carried out through a collaborative inquiry[1] with women in the village in 1998–2002. In the collaborative inquiry, I left the space open for the research questions to be decided by the whole group and for us to analyse our inquiry and the processes in the village together.

Social Nesting of Forest Governance

Decentralization of political power has been accompanied by the rhetoric of participation in management of resources (especially forests) and rural development. In a transfer of authority, questions about who constitutes the 'community' and the 'local' take on important connotations. It becomes incumbent on people to define themselves as a community and to define the issues of importance to them. In such a case, what are the issues that tend to get privileged and what impact does that have? I use the examples of the *Bruksha O'Jeevar Bandhu Parishad* (BOJBP; 'Friends of Trees and Living Beings') in Orissa's Nayagarh District and of the attempts at local forest management in a Swedish village to discuss what sustains and diminishes support for local efforts when, as a confluence of different factors, one issue becomes predominant in the community.

Friends of Trees crystallized as a network of village forest committees when men from several villages that had been protecting forests around their villages for some years came together in 1982. Eventually, the number of villages in the network increased and the movement grew to encompass the entire district. A larger federation called the *Nayagarh Jungle Suraksha Mahasangha* ('Nayagarh Forest Protection Federation'), formed in 1992, became the overarching organizational form for the various smaller networks. The forest cover in the district witnessed dramatic improvement in these years: forests began to grow back, streams that had dried up began to flow again and wild animals were once more sighted in the forests. In many ways, the success of the movement in its early years depended on the ability of its leaders to take on what they called a 'total approach to development'. Narayan Hazari, one of the founders, describes it as 'development in its economic, social, political, cultural, humanitarian, moral and spiritual aspects. Apart from economic reconstruction, the movement … worked for social reform, family planning, employment generation, removal of untouchability' (Hazari and Hazari, 1990). The religious and spiritual idiom was strong in the movement and appealed to the people's sentiments.[2] Through music and theatre and the written word, men and women communicated their message of environmental protection and social change at weddings, festivals, religious gatherings and other occasions.

It was also at these occasions that they spoke of the need for 'women's uplift'. Their drive, at least among the leaders, was accompanied by their effort to give up their caste membership. All this earned the movement support from many women and other groups such as the *Dalit* ('lower caste') community, who supported the movement even though they were not involved in the decision-making. Women were not involved in the decision-making either, but they were active in Friends of Trees campaigns and activities, such as the *padayatras* ('foot marches') in addition to day-to-day care of the families and the village during these occasions. Several were wives and relatives of Friends of Trees leaders. According to Narayan Hazari, when Friends of Trees went on campaigns, 'it was almost always the women who moved first'.

However, as the movement began to formalize and came in contact with outside NGOs and development agencies, the environmental agenda took centre stage, overshadowing the social agenda. A small amount of funding was provided by Oxfam, mainly for administrative costs. A 1992 evaluation commissioned by the Overseas Development Agency (precursor to the British Department for International Development) pointed out the lack of women in the organization, and funds were provided for a women's programme. The programme was meant as a capacity-building measure to enable women to be more active in the male-dominated forest organization. What is striking in the report, however, is that although it noted the absence of women from Friends of Trees management and leadership, little attention was given to the women who were already in the movement – those who organized and took part in the various functions and worked with the men. Initial training in tailoring and family planning was gradually replaced by support for savings and credit groups. Several of the women's groups that I refer to here were formed in the course of this programme. These groups were called the *mahila paribeshya vahinis* ('the women's environmental brigades'). As one of the *Mahasangha* staff told me, the groups were a tool to organize the women and integrate them into the forest movement.

But interviews with women revealed that the women's groups that had been most active and successful in bringing about change in their villages were those that had gone beyond the programme framework and had taken up a range of issues in their villages. In Nayagarh, women related differently to the forest and that depended in part on their social positions. Women from some villages collected fuelwood, others used *sal* leaves and other non-timber forest products, or collected seeds and other produce from the forests. Apart from mobilizing local savings in their groups, the women discussed village problems and worked on planting trees in the villages and on the outskirts of the forests, cleaning and maintaining village spaces, tending backyard plantations, and running small businesses together, such as making snacks. Some of the groups looked after herb gardens that they had set up. Several groups challenged local elites, such as male landowners, and others had resisted oppression by the police when the women sought to protect the forests. In such instances they made sure that social services due to the community were maintained, and by advocating community rights, they held government officials such as the block district officer accountable to the community. Through their work in the village and the forests, the women contributed to a more enabling environment for the management of the forests.

According to the women, issues of village management were inseparable from the forests. To be able to work with the forests, the committees and the federation also needed to deal with the concerns of the women in relation to the forests and also with other village activities. How these activities affected forest management – and how forest management affected them – had to be recognized, since the women had little interest in complying with forest rules and regulations that did not take into account their lived realities.

This aspect was even more apparent in the very different context of rural northwestern Sweden. In 1995, some men and women had formed a village association with hopes of reviving their area, which had suffered depopulation over the years. Important issues on their agenda were housing, schooling, childcare and working with the surrounding forests, which were part nature reserve and partly managed by a state-owned company for commercial purposes. The men heading the village association linked up with development practitioners from outside to work on the forest issue. The involvement of the outsiders shifted the focus of the village association. Attention to the single issue of economic benefits from the forests, in terms of employment for village men, came to dominate their efforts; other issues tended to get crowded out, excluding those who had other concerns. In 1998, several women in the village spoke to me about linking forest management to everyday village life such as the running of the village shop, day-care and old people's homes. As one woman put it, 'We cannot merely speak of managing the forests. The whole village needs to take part to realize the larger vision, the self-managing village.' To her it seemed incongruous to speak about 'local' management of the forests when many women felt that the village itself needed cohesive management. Within the women's forum, they spoke about the importance of nurturing community spirit to be able to work with village development and local forest management. They emphasized that the village association had originally taken up the issue of the forests as part of a larger vision of a 'living countryside'.

In other words, a locally managed forest would be difficult to achieve without a self-managed village where the villagers felt involved. Conversely, the forests were important for many reasons, and villagers hoped to be able to sustain the community by gaining some control over the forests. The women defined the forests as a social place that was integral to everyday community life.

Separation of the forest issue from other issues alienated many people in both study sites. Especially in Sweden, but also in Nayagarh, the nesting of the forests in other concerns was central to how people chose to live their lives. The forests were embedded in local and contemporary politics as well as in the history of the men and women. Forests thus need to be seen in a social context – as a 'social forest' encompassing not only the trees but also the ways in which the people relate to the forests and their relationships to each other.

By arguing for the need to take up other issues in relation to the forests, the women in both places sought to disturb the hierarchy of work as defined by the men (the forests being the big question that needed to be resolved right away versus the smaller questions concerning village development). The inclusion of the everyday and the spiritual and the interlinking of development needs made the Friends of Trees movement successful in its early days. Although explicit references to the word gender do not occur in its reports until 1988, this issue was taken up (at least in rhetoric – as women's uplift and problems of dowry) as the volunteers went on their *padayatras* from village to village. In the 1990s, several women organized women's *padayatras* and discussed women's situations as part of the environmental struggle.

In both places, local management could not be separated from cohesive village management. And for that, the terms and the agenda of the institutions for management needed to be negotiable; different ways of relating to the forests needed to be recognized. 'Local' management plans defined by a group of people (many of them outsiders) were limiting, especially for the women, who found it difficult to relate to local forest management if it did not take account of their concerns. The women's attempt to link the forests to other issues in everyday village life and to view them in a wider perspective also brought into focus the gender and power relations of the community where some issues were considered more important than others. A similar argument may be made about decentralizing authority: if the new decentralized organizations reproduce single-issue bureaucratic institutions for resource management, albeit at the local level, democratic and effective change is unlikely.[3]

Unintended Effects of Inclusion

In this section, I examine how the inclusion of women in committees and organizations does not necessarily benefit them and sometimes is used to preserve the status quo. I explain the circumstances that led to the reluctance of many women to attend meetings or show an interest in forest issues as defined by the village organizations. I discuss the forms of the meetings that made it difficult for women to be heard and discuss possibilities for change. Lastly, I suggest that it is important to accommodate different ways of organizing in the local community if women and men outside the formal organizations are to have any say in decision-making.

Disincentives to join, incentives to work separately

To ensure a democratic basis for decentralization, different groups are often invited to join decision-making committees and federations for forest management. In both case study areas, attempts to include women were made, especially in Sweden. Yet in both places, several women organized themselves and chose to work from within their own collectives. Both in Nayagarh and in Sweden, the women's collectives addressed a range of village activities and local development issues. Some of the women's groups in Nayagarh also became the forum where women could take up disputes in the home. They began to address questions of violence against women and what they called women's rights (as opposed to 'women's uplift') and issues of dowry. They spoke about the need for women to have their own federation in addition to the *Mahasangha* because they felt that their issues and approaches would not be taken up in the male-dominated forest federations.

The women's efforts to organize their own forum in Sweden and the Nayagarh women's wish to form a federation of their own were actively resisted or disregarded by several men in the formal organizations. Especially in Sweden, the women's forum was considered a problem for village harmony and a threat to the authority of the village association. This contributed to tensions in the village and the forest struggle. The women's forums were said to be different and not comparable to the male-dominated organizations. For example, in Nayagarh, the women's groups were seen as the space where women discussed 'women's work', saved money and 'got training'. Lister (1997, p147) writes that the invisibility of women's political activism is in part a reflection of the tendency to define politics within the narrow terms of the masculine sphere of formal politics. Some men in the associations sought to maintain these boundaries and in part drew their authority from their collaboration with the outsiders. In Nayagarh, some men in the *Mahasangha* said that one strong federation could take care of the women's problems as well, once some of the forest problems were resolved; as one man put it, 'Once the *Mahasangha* is stronger we can support the women in what they want to do. Without guidelines or ideas there is no point in women gathering. They need the *Mahasangha*'s support.'

The women's groups disagreed. As one woman told me, 'We can't keep waiting for them to take up our issues.' By taking care of several village problems, the women were in fact relieving the forest organizations of complications that could hamper their work with the forests. Also, as one of the field organizers of the women's groups wrote in a report, 'There is no point in talking about the environment if women have no power themselves. They cannot join a movement without looking at themselves and doing something about their own potential.' For the women's groups in Nayagarh, the freedom of being able to take up a range of questions and take action proved crucial in advancing their interests. The women's groups in some villages thus provided one possible space where the women could work to change their everyday reality, hand in hand with their involvement in forest activities.

Environmental relations are social and political relations. Problems arise often not from lack of forest wealth but from management decisions and discrimination. Studying the absence of women in forest user groups in Nepal, Lama and Buchy

(2004) write that some women were not interested in attending the meetings. The authors attribute this lack of interest in forming women's committees to women's illiteracy and inability to implement the rules according to the operational plan: in a process where women would have no influence, they were unlikely to be motivated to take part. Similarly, in Nayagarh, the rules and regulations drawn up by the male committees and outsiders were perhaps not ideal from the women's point of view. The women in Sweden were even less inclined to attend village association meetings focused on forestry and dominated by a few older men. These literate and capable women likewise had little incentive to participate if the forums did not accommodate their interests or give them a voice.

Agarwal (1997, p294) argues that women's forest protection groups could successfully be built on women's social networks. Although this may give the women some leverage vis-à-vis their villages, the expectation that women will be motivated to form groups for forest protection alone is questionable in light of their desire to take up a range of related issues and the discrimination they experience in their everyday lives.

Matters of form

One constraint for women involved the form in which meetings were conducted. In Sweden, on the face of it, nothing prevented the women from coming to village meetings, yet very few attended. According to some of them, the agenda was always defined by the men in the association and limited to certain aspects of the forests. Although the formal agenda allowed for other questions, this was always the last item, when there was little time left and the women had to go home to their children. An account from Nayagarh (Box 4.1) illustrates the need for different ways of organizing meetings. The form of the meeting in Maacchipada, for example, did not enable the women to take part and speak out as they did in Katrajhari. Societal restrictions on speaking out in the presence of elder male relatives and uneven power relations between men and women still come in the way of participation. Women's presence at meetings does not necessarily mean that they can participate.

Formal meeting procedures can open space for women's voices, as well as for men who normally would not speak. In the absence of opportunities to submit agenda items and sign up for a time to speak, as in Katrajhari (Box 4.1), meetings tend to be dominated by the elites. Researchers studying recruitment for managerial positions in firms have shown that more formalized procedures and open recruitment practices are associated with a greater share of management jobs held by women. Since informal channels and networks are normally dominated by men, formalizing personnel practices reduces men's share of management jobs (Reskin and McBrier, 2000). Unless decentralization efforts consciously build structures that enable different groups to speak, existing networks as well as informal social contacts that privilege the elites may self-perpetuate, and local

Box 4.1 A facilitating voice?

A district-level meeting was organized at Maacchipada by Vasundhara and Sanhati (an NGO alliance based in Bhubaneswar) to discuss community forestry. Several forest committees and *Mahasangha* members took part. About 45 participants, of whom only two were women, came from various villages. Throughout the meeting the two women sat and listened but did not say anything. I met them the next day in their village and asked them why they had gone.

They said they knew people from outside (the NGOs) would be there, and they wanted to tell them that women had been doing good work in their villages. But they were not given a chance to speak. A *Mahasangha* staff member told them that if they had just raised their hands, they would have been asked to speak; the women replied that the long speeches made by the men made that difficult. 'It isn't that easy for women to speak out. Nobody asked me to speak, I was waiting to be asked. It's only when we become more than half in numbers at such meetings that we will be able to speak out.'

At another *Mahasangha* meeting, at Katrajhari, one of the women actually did speak. Her participation had been facilitated by the form of the meeting: since there were so many attendees, the organizers reserved slots for those who wanted to speak, and participants could submit their names. She was then assigned a slot and could speak uninterrupted.

management will mean only that oppressive authority over the forests, once exercised by a distant government, is now closer to home.

Opportunity to speak is one thing; being heard is another, and that depends very much on local relations of gender and power. The question, as Spivak (1990, p59) writes, is not 'who should speak but who will listen'. Several women in Nayagarh said they could participate in forest federation meetings outside their villages but not in their own villages, where older male relatives were present. In these village meetings, women would have little choice but to legitimize decisions that might not reflect their priorities or might even be contrary to their own interests. The willingness of outsiders to solicit and listen to different voices at meetings is thus important.

Critical mass and critical action

The women of Maacchipada needed 'to be more than half in numbers' if they were to have an impact on resource management (Box 4.1). In academic literature, some feminists have argued that a critical mass of women is essential to bring about change that is in their favour (for example Agarwal, 1997). But how can women reach a critical mass if power structures and gender inequality are unchanged? What is needed is 'critical action' (Dahlerup, 1987). In both study sites, women chose to organize their own groups in addition to joining the forest committees and associations, which they felt had no place for their concerns. In their own spaces they could get strength from one another and take action.

Attempts to include women within the formal organizations, ostensibly an effort to broaden the democratic base, can in fact be seen as a means to maintain the gender status quo. Both in Nayagarh and in Sweden, the men wanted the women to join the village associations. But making a space for women did not include changing the decision-making structures and procedures that disadvantaged women. Although both the forest organizations and the women's groups agreed on the need for local resource management, the real issue was whose definitions of local management would be accepted, on whose terms activities would be conducted, and who represented the community – the 'local'. Decentralization efforts thus have to confront not only struggles over resources but also the 'struggle over meanings' (Peters, 1984).

The case studies show that women and other marginalized groups may need to participate through a variety of forms – in concert with each other, from within their own spaces and also as individuals within formal institutions. The women's different ways of working were not accommodated by existing organizations or by the new institutions for forest management, which all too soon began to replicate the bureaucratic norms that they sought to replace. Local management of forests is thus intertwined with questions of power and social relations. And in these issues outsiders play a crucial role, as I show in the following section.

Skewing Effects of Outsiders

Decentralization is not only a top–down process by the state, and local management is not only a community effort. Local management is defined in collaboration between the villagers and outside actors. The boundaries between inside and outside are not always clear, however. In this section, I look at the roles played by outsiders and the implications for field practitioners.

Conferring legitimacy

In Sweden, the development activists and practitioners working with the village association strengthened the position of the men in the association, thereby further legitimizing and enhancing their power in the village. Criticism by some women of the undemocratic methods of the men working with the forest project was dismissed as village politics – something to be resolved by the villagers themselves, a problem that outside agents should not engage in. However, by supporting the men in the association and assuming that their ideas were representative of the entire community, the outsiders engaged directly in village politics, thus affecting power relationships in the village. By separating the question of the forests from the other parts of the village agenda (an approach attributable to their forestry education, their profession and an official sectoral approach), the development practitioners concurred with the men. They disregarded the women's critique of the forest project as personal politics and thereby legitimized the forest organization's failure to reach out to the rest of the village or accept criticism. The result: the local forest management project did not have the support of all the villagers, especially the women.

Similarly, my own involvement in the village had the effect of bringing to the surface disagreements about the way local management was to be organized. Thereafter, the collaborative inquiry approach, by allowing the women to direct the course of the research, brought to light other forest-related issues and focused on the importance of community spirit and an environment that would enable them to participate in equitable and sustainable forest management.

Nayagarh already had a vibrant community forestry movement when the outside agency began to collaborate with the village. Realizing that women were absent from the formal structures, Oxfam pushed for a women's programme, but the intervention was not made in consultation with the women whom they were trying to help. The programme was planned by Oxfam staff, initially with Friends of Trees, and addressed tailoring, family planning, savings and credit; Oxfam channelled funding for the women's programme through Friends of Trees. The women insisted that Friends of Trees leaders take a more inclusive approach to decision-making, and a conflict arose. Oxfam was by now more interested in supporting the *Mahasangha*, which it saw replacing Friends of Trees as the people's organization. After trying to mediate between the women and Friends of Trees, Oxfam withdrew its funding and, by implication, its official support. The women's groups thus got caught in the discord between Oxfam and Friends of Trees. This was unfortunate for the women, who were in the process of building up a network and working actively on village development. It was not so much the funding (which was quite modest), but the official support that was of importance for the women's groups. Thus, on the one hand, problems and inequalities were recognized by the outside agency, but on the other, given their mandate to work with 'community organizations', Oxfam felt that there was little they could do to intervene or to support the women's groups and their staff. Quite obviously the male committees and the BOJBP were taken to be representative of the entire community and the women's groups were not. Unlike the case in Sweden, Oxfam acknowledged gender inequalities but felt unable to do very much about them. Ultimately, however, Oxfam sought to open up the *Mahasangha* to include more women.

Supporting alternative spaces

In addition to having a place in the formal arenas, it was important for the women to have their own space that also enabled them to take action and gave them self-esteem. The women's groups supported by Friends of Trees with Oxfam funding provided such an opportunity, and official support gave them legitimacy. The meeting in Maacchipada (see Box 4.1) is one example of women's attempts to gain legitimacy. Whatever the relevance of the women's tree-planting activities for a discussion on a comprehensive forest policy, it was important for them to be recognized by outsiders and be acknowledged for their contributions to the forest management efforts.

Creating alternative spaces outside mainstream organizations is not without complications, however. The women from a *dalit* village in Jakalla that was not a part of Friends of Trees or the *Mahasangha* area created an informal market: they made and sold bamboo goods to women in other villages. One woman told me:

> *We go far and sell our goods ... rice huskers, baskets, nets ... to other women. ... Only the women buy these things since they know the quality. That is why we have to go from village to village instead of going to the market. ... Even the men would like to purchase because they don't want the women to spend the money for these things. ... We exchange the goods for rice, wheat, potatoes, onions ... everything. We do not go to the weekly haat in Gania, Daspalla ... because rice is better to get. ... We can't manage all the bargaining, manipulation. ... Other villagers sell such things in the market ... those close to the market.*

For the women, this mutually beneficial relationship had its costs, since they had to travel, leaving their small children in the care of older siblings. The opening up of local *haats* (markets) where women could sell their non-timber forest products and other goods gives them important leverage, but as this case indicates, as soon as women do enter the markets, they come under the conditions set by others and over which they have little control. In establishing women-to-women channels, retaining the right to be the best judge of the quality, and exchanging in kind, the women of Jakalla and their customers maintained control over their exchange. The informal market, however, did not involve cash, which is an important source of power.

This situation has striking parallels with decentralization efforts that seek to include women in mainstream organizations. Women's informal groups – places where they have a certain measure of power – may need to be supported even as male-dominated organizations are opened to them, and the links between formal and informal modes of political participation may need to be strengthened.

I suggest that bringing women into existing structures is not always beneficial for them and may in fact be a step backwards. In the study sites in both Sweden and India, the outsiders tried to include more women in the formal organizations. The story of the women in Jakalla shows what can happen. It would benefit the women to be part of mainstream organizations, but is that enough? What about women's desire to organize separately? The women's efforts to organize themselves caused consternation among the male committee leaders.

Discussion

Local women's groups are not often regarded as competent or valid partners in official policy development. Women's organizing in the countryside often takes place informally and outside the big programmes or official life, a point discussed by researchers both in the North and the South (for example Rönnblom, 1997; Purushottaman, 1998). In an analysis of collective action in the forests, Agarwal (2000) distinguishes women's involvement in resource management, which tends toward agitation, from the more long-term and cooperative organizations that are dominated by the men. From my case studies it appears that women's networks and actions might seem sporadic and provocative because their cooperative work in the villages often remains invisible for outsiders focused on a single issue, such as the forests. The women tended to organize in response to specific problems that involved different issues.

One such instance took place in the village of Hariharpur when the women came together to protect the forests from intruders. This 'spontaneous' action had antecedents in women's networks that had formed around other issues. But their organizing does not assume the recognizable, more permanent forms that are considered viable organizations within development work. In academic theorizing as well as in policy approaches, long-term work with the forests tends to be prioritized over immediate, everyday concerns, which may change but which may be crucial for day-to-day community life and the ability to work with the forests. Although disciplinary or sectoral divisions may make theory building or policymaking simpler, the issues remain inseparable on the ground, and the categories make it difficult to grasp the interconnection between management of a resource and the issues and social structures of the people dealing with them.

Neither in India nor in Sweden did the women organize themselves solely around forest issues. But in many ways the issues were implicitly connected, and forest management could not be separated from its broader context. A broad approach that takes into account the social context is needed to create an enabling environment for forest issues as well as for other community development matters. To be effective, community forestry needs to be embedded in community life.

Access of the women to the outside world is often mediated by male leaders and organizations. By organizing themselves with the help of the programme and then taking up issues outside the formal programme, the women had challenged this way of working as well as male authority. Outside agencies need to recognize the potential for local development and resource management that informal groups represent. While Oxfam wanted to support the women's organizing, it felt unable to do so since there were no structures in place that would allow them to support the women without channelling support through the men. In Sweden, outside intervention inadvertently legitimized existing relations in the village association, thereby cementing unequal practices. The increasing bureaucratization of village development work and local management meant that the women were required to accommodate themselves to existing norms and structures rather than allowing for the structures to be changed to accommodate the women's standpoints. Paradoxically, this understanding of collaboration was reinforced by development workers who were trying to promote democracy and greater equality in decision-making.

Women are often invited to join committees and political organizations because it is assumed that they have different views on issues and can represent different constituencies. However, women, no less than men, have different interests arising from their economic and political relations (see Pringle, 1997). Both in Sweden and in India, women's groups formed in response to questions of power rather than to express different views based on their sex. The assumption is often made that the formation of a women's group automatically leads to women's empowerment. It is, rather, the structural constraints that need to be recognized. Women and other marginalized groups need to be able to participate in a variety of ways, both separately and within formal organizations, and to argue their positions without the assumption that being a woman makes a person think in a particular way or that a few women can represent all other women.

Neither does dialogue solve all problems, given that language competence and the ability to speak out are not shared equally. Kohn writes that a discursive vision of politics – a politics based on the assumption that dialogue is needed to reach consensus – tends to reproduce the status quo. She points to empirical evidence that political struggles that take place on the basis of deliberation are heavily weighted in favour of the elites. What is needed, she argues, is another definition of politics, rooted in contestation, struggle and resistance, and structural or institutional changes in the basis of power, which may even require strategically separate spaces – what Fraser calls 'subaltern counterpublics' (cited in Kohn, 2000). The women's groups in both Sweden and India provided just that alternative space.

Decentralization provides an opportunity to create new structures for managing forests and village relationships that accommodate different ways of relating to the forests. The forms in which people choose to participate need to be flexible and encompass both formal and informal organizations. What is needed, as Lister (1997, p163) argues, is a reconstruction between formal and informal political institutions so that the former can be better grounded in and more accountable to the latter.

Processes of development and local management take place in a gendered terrain. Power and rights to the people mean little if one does not take into account gendered differences, especially in resource management and its masculine ideals of the male forester, the male fisherman, the male hunter. Otherwise, one may be tempted to ask whether decentralization, devolution and local management are an ideology whose time has come for men – but not for women.

Notes

1 Collaborative inquiry is a form of participatory action research. See Reason (1994) for examples and guiding principles.
2 Human and Pattanaik's (2000) history of the movement also emphasizes this point.
3 In Arora-Jonsson (2005) I discuss at greater length how these local organizations tended to reproduce the bureaucratic structures that they sought to replace.

References

Agarwal, B. (1997) 'Environmental action, gender equity and women's participation', *Development and Change,* vol 28, no 1, pp1–44

Agarwal, B. (2000) 'Conceptualising environmental collective action: Why gender matters', *Cambridge Journal of Economics*, vol 24, pp284–310

Arora-Jonsson, S. (2005) 'Unsettling the order: Gendered subjects and grassroots activism in two forest communities', PhD thesis, Swedish University of Agricultural Sciences, Uppsala, Sweden

Dahlerup, D. (1987) 'Confusing concepts – Confusing reality: A theoretical discussion of the patriarchal state', in A. S. Sasson (ed) *Women and the State: The Shifting Boundaries of Public and Private*, Routledge, London/New York

Hazari, N. and Hazari, S. C. (1990) 'Environmental revolution in Orissa: The Buddhagram experiment', in R. Akhtar (ed) *New World Environment Series IV – Environmental Pollution and Health Problems*, Ashish Publishing House, New Delhi

Human, J. and Pattanaik, M. (2000) *Community Forest Management: A Casebook from India*, Oxfam Publication, Oxford, UK

Kohn, M. (2000) 'Language, power, and persuasion: Towards a critique of deliberative democracy', *Constellations*, vol 7, pp408–429

Lama, A. and Buchy, M. (2004) 'Gender, class, caste and participation: Community forestry in Central Nepal', in S. Krishna (ed) *Livelihoods and Gender: Equity in Community Resource Management*, Sage Publications, New Delhi/Thousand Oaks/London

Lister, R. (1997) *Citizenship: Feminist Perspectives*, Macmillan, Basingstoke/London

Peters, P. E. (1984) 'Struggles over water, struggles over meaning: Cattle, water and the state in Botswana', *Africa,* vol 54, pp29–49

Pringle, R. (1997) 'Feminist theory and the world of the social', in G. Murray and G. Tulloch (eds) *Feminism in the 1990s*, Sage, London/Thousand Oaks/New Delhi

Purushottaman, S. (1998) *The Empowerment of Women in India: Grassroots Women's Networks and the State*, Sage Publications, New Delhi/Thousand Oaks/London

Reason, P. (1994) *Participation in Human Inquiry*, Sage, London
Reskin, B. F., and McBrier, D. B. (2000) 'Why not ascription? Organizations' employment of male and female managers', *American Sociological Review*, vol 56, pp210–233
Rönnblom, M. (1997) 'Local women's projects', in G. Gustafsson (ed) *Towards a New Democratic Order? Women's Organizing in Sweden in the 1990s*, Publica, Stockholm
Shackleton, S., Campbell, B. M., Wollenberg, E. and Edmunds, D. (2002) *Devolution and Community-Based Natural Resource Management: Creating Space for Local People to Participate and Benefit?* Natural Resource Perspectives 76, Overseas Development Institute, London
Spivak, G. C. (1990) 'Poststructuralism, marginality, postcoloniality and value' in P. Collier and H. Geyer-Ryan (eds) *Literary Theory Today*, Polity Press, Cambridge, UK
World Bank (2000) *World Development Report 2000/2001: Attacking Poverty*, World Bank, New York

5

Community Forestry in Nepal: Decentralized Forest Governance

Ganga Ram Dahal and Apsara Chapagain

Nepal's community forestry has become an example of progressive legislation and policies in the decentralization of forest management. It has attracted international attention because in Nepal decentralization is linked with emerging issues – sustainable forest management, forest governance, policy advocacy, equity, gender, poverty and the role of civil society in community forestry. In particular, the role of the forest user group network in legal advocacy, capacity-building and the establishment of democratic governance on a wider scale shows the unique strength of the community forestry approach in Nepal. Since the enactment of the Forest Act in 1993, the government of Nepal has gradually been handing over parcels of national forest, particularly in the mid-hills, to local communities based on a forest management plan agreed between the District Forest Office and local forest user groups. As of December 2005, some 1,185,565ha of forest had been handed over to 14,227 forest user groups for them to protect, manage and utilize (Department of Forests, 2005).

Despite many positive outcomes, however, community forestry in Nepal has faced mounting challenges, limitations and shortcomings, particularly in implementation. Notable challenges include elite capture, inability to provide significant contribution to livelihoods, the persistence of poor and disadvantaged sectors in the community, and imbalances between social needs and the environmental agenda. This chapter presents an overview of the decentralization process, institutional arrangements, achievements and challenges for community forestry in Nepal.

Milestones of Decentralized Forest Governance in Nepal

Many countries in Asia and the Pacific that have adopted decentralization policies in forestry seek to achieve effective management of forest resources by ensuring sustainability on the one hand and supporting local people's agendas on the other (Fisher, 1999; Agrawal and Ostrom, 2001; Ribot, 2004; Capistrano and Colfer, 2005). Based on these broader rationales for decentralization in the forestry sector, Nepal has also initiated efforts toward decentralization of forest management via the 'roadmap' in the Master Plan for the Forestry Sector of 1988 (Hobley, 1996).

The following acts and regulations are considered milestones in the history of decentralized forest governance in Nepal. Based on these legal foundations, the approach of community forestry has been evolving over the last 28 years. Some of the features of the policies, acts, rules and regulations are briefly discussed below.

Decentralization Act of 1982

To overcome the continued failure of the centralized approach and the *panchayat* (literally, 'assembly of five', referring to the people's representatives at the local level) approach to decentralized rural development efforts, the government of Nepal passed the Decentralization Act in 1982. This act formalized the duties and responsibilities of village *panchayats* and ward committees. All development interventions were required to adopt the 'user group' concept when implementing projects at the local level. The idea behind this approach was to ensure local participation, establish linkages between local and national planning, mobilize local resources, and strengthen local institutions for development in the long run.

Master Plan for the Forestry Sector of 1988

The development of the Master Plan for the Forestry Sector was a response to the perceived need for re-examination of the forestry sector in Nepal. The Master Plan provides the main basis and framework for developing policies for managing the country's forest sector. It aims to mobilize, conserve and manage forest resources in a sustainable way and thereby maintain a balance in the demand for and supply of forest products, create income and employment opportunities within the sector for poor and marginalized households, promote people's participation, enhance productivity, and develop appropriate land-use plans. The Master Plan comprises 12 programmes, of which the private and community forestry programmes are given the highest priority. All the accessible forests in the mid-hills were to be handed over to local communities with the formation of user groups. This plan opened up avenues for participation by local people in the management of forest resources. However, it did not address the lower-lying, more valuable forests of the *terai* (in southern Nepal) or the high-altitude forests.

Forest Act of 1993 and Forest Regulation of 1995

These acts, a breakthrough for community forestry in Nepal, provide a legal basis for the implementation of community forestry and build on the Master Plan for the Forestry Sector of 1988. As envisaged in the Decentralization Act of 1982, the Forest Act of 1993 recognized forest user groups as legal entities and acknowledged five categories of national forest: community forest, leasehold forest, government-managed forest, religious forest and protected forest. Major goals of this act were to meet the basic needs of local people, attain economic and social development, promote a healthy environment, promote development and conservation of forests and forest products by managing national forests, and provide assistance for the conservation and development of private forests. Despite some progressive policy provisions toward decentralization, however, the Forest Act of 1993 has several gaps and inconsistencies. For example, forest user groups are given only usufruct rights; forest ownership is retained by the state. Similarly, there is a lack of clarity about management based on the size of forests and about the roles of different actors. Some of the major highlights of the provisions of the Forest Act of 1993 and the Forest Regulation of 1995 are listed in Box 5.1.

Box 5.1 Forest Act of 1993 and Forest Regulation of 1995

Whereas it is expedient to meet the basic needs of the public in general to attain social and economic development and to promote a healthy environment and to ensure the development and conservation of forest and the proper utilization of forest products. (Forest Act, Preamble)

The users relating to any forest desirous to utilize forest products by developing and conserving such Forest for the collective interest may constitute the Users' Group As Prescribed. (Forest Act, Section 41)

For the registration of the Users' Group ... an application shall be submitted to the District Forest Officer in the Prescribed form along with the constitution of the Users' Group. ... the District Forest Officer shall ... be given the certificate of registration. ... The Users' Group ... shall be an autonomous and corporate body having perpetual secession. (Forest Act, Sections 42 and 43)

While preparing a Constitution ... and Constituting a Users' Group according to the Constitution, action shall be taken on the basis of consensus rather than considering geographic boundaries of wards, villages, towns and districts. (Forest Regulation, Rule 29(4))

. . . the District Forest Officer shall depute as soon as possible a technical employee to the concerned area for the purpose of providing technical and other co-operation. . . . The District Forest Officer shall provide technical and other co-operation required by the concerned Users' Group to prepare a Work Plan . . . (Forest Regulation, Rules 27(2) and 28(3))

Forest Sector Policy 2000

Unlike other policies and acts, the Forest Sector Policy of 2000 reverted to the conservation agenda and made it obligatory for the community forestry user groups to pay 40 per cent of their earnings from timber sale to the government (Kanel, 2006). Many consider this a government decision with adverse implications for forest decentralization in Nepal, as it curtails the authority devolved to the local communities.

Institutional Arrangements to Implement Community Forestry

Primary actors in the implementation of community forestry are the national, regional, district and range-post levels of the Department of Forests and community groups (forest user groups). These groups have established their own network, the Federation of Community Forestry Users of Nepal (FECOFUN) at national, regional, district and range-post levels (Figure 5.1).

Government organizations

The Department of Forests was established in 1942 with the primary role of protection and management of the national forests. Since 1942, it has undergone a series of restructuring processes in order to make the organization compatible with the changing context. However, still the institutional set-up remains unusually hierarchical, particularly in the traditional attitudes of government bureaucrats and the slow process of transformation of the role of the foresters. This does not support the decentralization process in many cases. At present, there are 5 Regional Forest Directorates (responsible for coordinating, planning and monitoring district forestry programmes), 74 district forest offices (responsible for planning implementation at the district level), 92 *Ilakas* (sub-districts) forest offices and 698 range posts. After the adoption of the community forestry concept, the government forestry personnel gradually shifted their role from policing and control

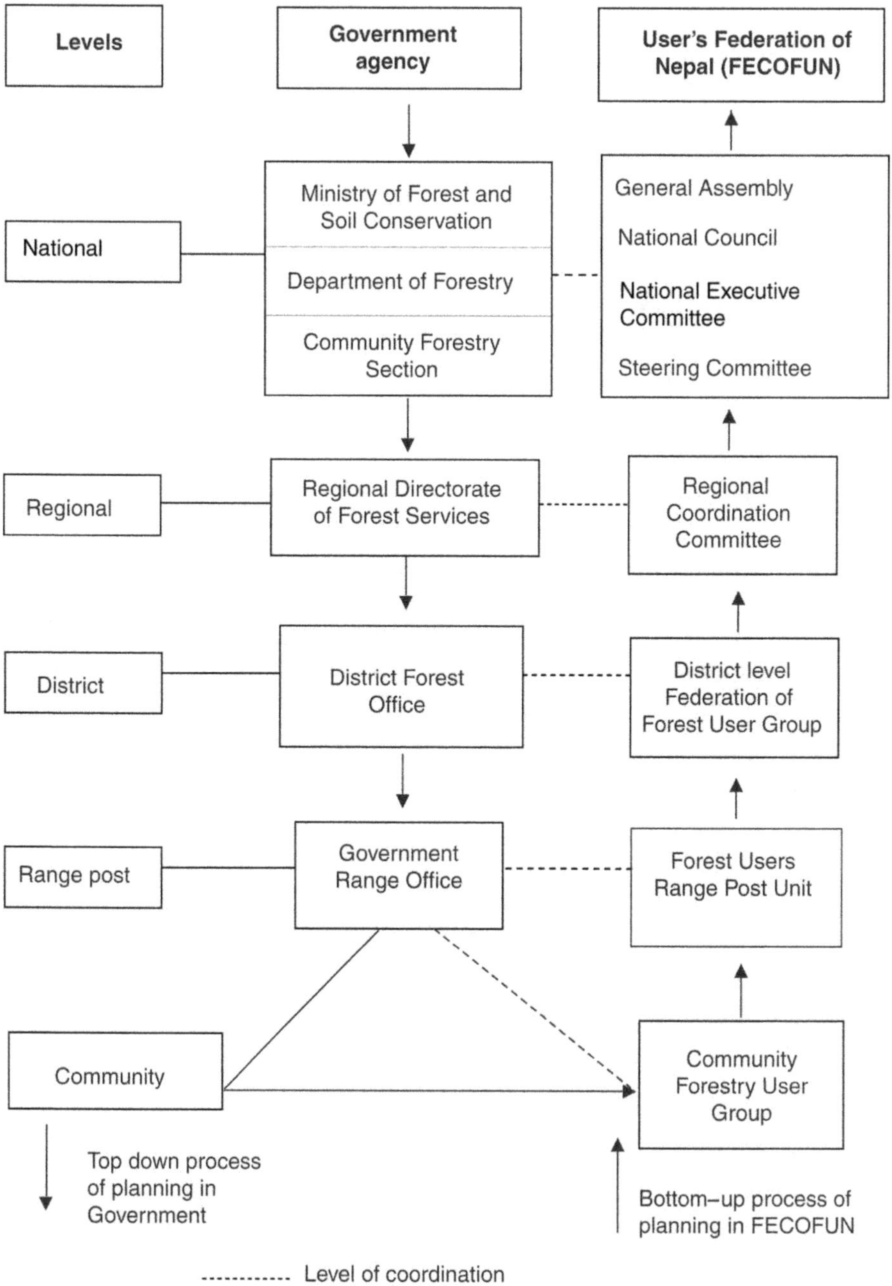

Figure 5.1 *Organizational structure of Ministry of Forestry and FECOFUN*

to extension work and facilitating. However, still there are many senior bureaucrats who remain committed to the command-and-control role and are not happy in their new one. In some cases, such bureaucrats seek to retain the devolved power through manipulation and amendment of the existing Forest Act and legislation. However, Nepal's strong civil society networks such as the users' federation have interfered with these efforts. For example, the government decision to give *terai* forest management to an international company from Finland was altered due to continued objection and pressure from the federation of forest users.

Forest user groups

The Decentralization Act introduced the concept of forest user groups in 1987. The Master Plan for the Forestry Sector of 1988 and the Forest Act of 1993 formalized the concept and gave a legal basis for the groups to function as autonomous institutions in the management of forest resources. The following are some of the legal provisions that enabled forest user groups to build a strong institutional basis:

- The user group shall be an autonomous and corporate body having perpetual succession (Section 43(1)). It shall have a separate seal of its own (Section 43(2)). The user group shall have a separate fund of its own (Section 45(1)). The user group as a legal person may acquire, possess or transfer or otherwise manage movable and immovable property (Section 43(3)).
- The user group shall have to prepare a work plan for the community forest (Rule 28(1)).
- The user group shall collect, sell and distribute the forest products which are available pursuant to the work plan (Rule 32(1)).

After the enactment of the Forest Act of 1993, the formation of forest user groups began throughout the country. By 2003, there were 12,725 forest user groups across Nepal, with more than 1,400,000 households managing 15 per cent of Nepal's total forest land area (Table 5.1). Under the Forest Act of 1993 and the

Table 5.1 *Forest user groups in Nepal*

Regions	*Forest user groups*	*Area (ha)*	*Households*	*Total income (Nepal rupees, thousands)*	*Total expenditures (Nepal rupees, thousands)*
Mid-hills	9353	696,044	976,715	85,112	43,407
High mountains	2456	189,843	248,619	10,070	4532
Terai	916	124,853	196,967	82,898	72,950
Total	12,725	1,010,740	1,422,301	178,080	120,889

Source: Department of Forests (2003)

Forest Regulation of 1995, forest user groups are allowed to find ways to achieve financial sustainability. This act requires that forest user groups spend a quarter of their income on forest management. However, many forest user groups spend most of their income on construction of local trails, school buildings, culverts and irrigation canals. Others are supporting livelihoods options for the rural poor and marginalized groups in their communities. The forest user groups are voluntarily united under the umbrella of the Federation of Community Forestry Users of Nepal (FECOFUN) to ensure their rights are protected and not curtailed by the government.

The Federation of Community Forestry Users of Nepal

FECOFUN, established in 1995, is a representative body of the community forest user groups and has the following aims:

- to advocate the rights of community forestry users;
- to strengthen local capacity;
- to establish linkages; and
- to lobby on behalf of the forest users.

FECOFUN is an autonomous, non-profit, membership organization that is accountable to its constituency – the forest user groups. Almost 5 million people are affiliated with the network, and it is one of the largest civil society networks in Nepal. FECOFUN works with non-governmental organizations and local and international donors to raise awareness in civil society about community forestry legislation and the roles and responsibilities of different actors in community forest management. It also promotes the empowerment of women and disadvantaged groups in the community by facilitating their integration into the community process and making sure that their voices are heard when decisions are made in the community. Furthermore, FECOFUN provides legal advice and assistance to forest user groups and collaborates with researchers, academics and other civil society groups to provide technical inputs for income-generating activities within forest user groups.

Besides acting as a watchdog for the rights of forest user groups, FECOFUN also engages in activities related to the democratic rights of citizens and the institutionalization of such processes. The federation became involved in Nepal's democratic movement in 2006, when it mobilized forest user group members in protest rallies and mass movements in favour of the re-establishment of democracy.

Although the idea to form a federation was originally facilitated by external donors, the institutional strengthening and expansion of the network has been managed primarily by FECOFUN itself. The federation has established a system for generating financial resources within the forest user groups at the local level, whereby a certain percentage of their contributions supports the national-level federation as well. This mechanism helps FECOFUN remain independent, thus facilitating its ability to engage in advocacy and lobbying. Even in policy dialogues with government teams, the federation has consistently spoken up clearly in favour of user groups and continued its advocacy until its demands were met. For example, FECOFUN played a significant role in persuading the government to expand community forestry policy from the mid-hills to the *terai*. Similarly, it has played a significant role in protesting against the government's decision to sign a contract with the Timber Company of Finland for the management of the valuable *terai* forest in Bara District. Despite its positive role in protecting the rights of the forest users, however, there are some internal institutional limitations within FECOFUN, such as the over-representation and influence of people belonging to a single political party in the executive committees at local, district and national levels.

Achievements of Community Forestry

As a result of devolving managerial rights to user groups, the community forestry programme has had noteworthy achievements, including forest restoration, social inclusion and representation, improvement of community infrastructure, rural development, and contributions to poverty reduction. Barren lands, denuded hills and degraded forest lands have been converted into productive woodlands. Lost greenery is now restored. Forest management by communities has contributed to environmental improvement, although the total contribution has not been quantified. With improved forest conditions, the availability of forest products, local people's rights of access and the supply of forest products to poorer households have increased (Gautam et al, 2004). As a result, the time women spend collecting fuelwood, one of their main tasks, has decreased. The contribution of community forestry to watershed protection, soil erosion control, protection and restoration of water sources, environmental purification, and a healthier living environment has been enormous, although additional scientific measurement is still required for real quantification. Besides achieving the physical target of handing over the national forest to local forest user groups, community forestry has simultaneously enhanced leadership and community development, democratic processes, poverty reduction, gender equity and social inclusion.

Leadership development

Each forest user group has formed its own executive committee. Meetings of the executive committee and the user group general assemblies are two important decision-making forums, and such organized meetings and assemblies are regular events. As a result of the frequent discussions and meetings, both men and women have developed their leadership skills as they make and implement decisions. A study carried out by FECOFUN in the west of Nepal showed that even disadvantaged, marginalized women and *Dalits* (members of the so-called 'untouchables' caste) have taken on major positions in some executive committees (Table 5.2). The empowerment programme in the forest user groups has gradually made such disadvantaged members aware of their rights, roles and responsibilities. The database of the Department of Forests also lists more than 721 forest user groups registered as women-only groups.

Table 5.2 *Representation of* Dalits *in forest user group executive committees*

District	*User groups studied*	*Committees without* Dalits	*Committees with* Dalits
Jumla	48	24	2
Dailekh	20	4	0
Surkhet	50	4	3
Dadeldhura	63	11	0
Baitadi	50	20	5
Total	231	63	10

Source: FECOFUN (2005)

Because of the development of leadership skills in community forestry, many leaders of forest user groups are elected to local political bodies, such as village and district development committees. Some forest user group leaders are also heading other local institutions, such as school management committees, drinking water groups, irrigation committees and clubs.

Women's leadership development has benefited from the community forestry process in Nepal. One of the studies carried out by FECOFUN in the Rapti zone in western Nepal in June 2005 (Table 5.3) found that women are not only members of forest user groups but also in positions to influence decision-making in community forestry.

Table 5.3 *Women in forest user group executive committees*

District	*User groups studied*	*Women chairs*	*Women vice chairs*	*Women secretaries*	*Women treasurers*	*Total women in executive committees*
Rolpa	39	14	17	12	13	153
Pyuthan	60	10	28	7	18	235
Dang	72	5	19	7	5	295

Source: FECOFUN (2005)

Establishment of democratic practices

There are ample opportunities within forest user groups to establish democratic practices. Many groups are trying to make decisions based on general consensus and prepare constitutions and operational plans in a transparent and participatory way. The rules and regulations are developed and approved in general assemblies. Following the procedure of general consensus, the users form an executive committee. In regular meetings, the executive committee and general assembly amend the constitutions and operational plans. In some cases users can openly criticize, debate, self-criticize and praise one another. The income and expenditures of the group are reported and discussed in the assemblies before being approved. Some forest user groups already practise public audits.

The principal power of forest user groups lies with the general assembly. As required, members may devolve executive power to the committee, but that can be withdrawn if the power is not properly used. In this way, members perform the role of cabinet, court and parliament. Forest user groups have also acted as local governance structures and delivered services when there was a political vacuum with no elected local bodies or parliament.

Community development activities

Besides investing their funds in forest management, many forest user groups have contributed to community development activities. Forest user groups have reformed the rural infrastructures in their own way. Using income from the sale and distribution of forest products and contributing their labour, forest user groups have constructed schools, roads, drinking water supplies, irrigation systems, electricity, health posts, public toilets, embankments, community buildings and soil erosion control systems. As a result, some public services previously missing in the local communities are now available to the general public. For example, Prajapati Forest User Group in Jhapa District in eastern Nepal built a connecting road to the village from the highway, and Sati Karnali Forest User Group has been operating an ambulance service in its area. Such initiatives have significantly contributed to community development, making forest user groups among the leading community development actors at the local level.

Contribution to poverty reduction

Helping reduce rural poverty is the main objective of community forestry, as stipulated in the Master Plan for the Forestry Sector (1988) and also in the 10th five-year plan. Many forest user groups have been trying to make their activities more beneficial to the poor and are writing provisions that address the issue into their constitutions and operational plans, including efforts to contribute to the livelihoods of the poor in the community, to safeguard poor people's rights, and to ensure their access to resources, decision-making and benefit sharing (Malla et al, 2003). Some forest user groups have been providing forest products free or at subsidized rates to members identified as poor. Many allocate a certain percentage of the funds from income-generating activities for the poor and support the schooling of poor children by providing scholarships, school uniforms and school supplies (Table 5.4). Some forest user groups have also supported the landless poor by providing land for them to construct their houses. These examples indicate the great potential for community forestry to address poverty at the local level.

Table 5.4 *Examples of forest user group programmes for the poor*

District	*Forest user group*	*Programmes for the poor*
Jhapa	Bans Bari	Providing forest products free of charge Scholarships for children Construction of 60 houses
Kailali	Gyansee	Support for four poor households in swine farming Cultivation of herbal plants for 16 poor households 25% of forest user group income set aside for the poor

Source: FECOFUN (2005)

A study by Pokharel and Nurse (2004) in Dolakha, Ramechhap and Okhaldhunga Districts in mid and eastern Nepal showed that forest user groups are becoming more responsible towards the poor. Sixty-two per cent of forest user groups have identified the poor households in their communities and have approved and included provisions focusing on the poor in their constitutions and operational plans. Interestingly, 95 per cent of forest user groups have provided forest products to the poor at subsidized rates, and 56 per cent have quotas for representation of women and *Dalits* in leadership roles (at least 30 per cent and 15 per cent respectively). Forty-nine per cent of forest user groups have been providing scholarships to the children of the poor and 42 per cent provide discount loans.

Gender and equity

A forest user group is principally an institution where people of diverse religion, caste, gender, class and strata can participate equally. Official policies promote social inclusion. But there are still many forest user groups that have not properly followed the principle of social inclusion. For the successful implementation of community forestry, a community's poor, women, *Dalits* and marginalized groups should participate meaningfully and equitably in decision-making. Those who previously participated only passively in forest and community development activities should have access to more active decision-making, planning, programme implementation, monitoring and evaluation, and benefit sharing – and risk sharing as well.

Table 5.5 shows women's representation on the executive committees of groups in the same districts. Only 24 per cent of women participate in a forest user group executive committee, which indicates that there is still much to do to achieve gender balance in the community forestry process in Nepal.

Table 5.5 *Women in forest user groups holding positions on executive committees*

District	*User groups studied*	*Chair*	*Vice chair*	*Secretary*	*Treasurer*
Jumla	48	15	13	3	10
Dailekh	20	2	2	2	2
Surkhet	50	7	12	7	8
Dadeldhura	63	25	30	21	24
Baitadi	50	4	11	4	5
Total	231	53	68	37	49

Source: FECOFUN (2005)

A study carried out by Springate-Baginski et al (2003) highlighted the role of cultural norms and caste in women's participation in community forestry. The study indicated that in *Brahmin* (so-called higher caste) communities, women's participation was strongly discouraged; cultural taboos and conservative social norms restrict women's open involvement in community activities. In *Tamang* (so-called lower caste) communities, in contrast, women's participation was more common; the culture is more liberal and open about women's involvement. In the same study, inequitable practices were found, such as favouritism in forest product distribution, exemption from fines for friends and relatives of executive committee members, membership fees unaffordable by the poor, and restrictions on charcoal making (which disproportionately affects the poor, many of whom depend on selling charcoal for their livelihood).

Challenges and Limitations in Community Forestry

Adoption of the decentralization policy in forestry has come about after a long journey along a rough road and its implementation is still selective. The system of decentralization in forestry started in Nepal when government was pursuing autocratic goals and governance was highly centralized. Until restoration of democracy in 1990, the decentralization policy was merely a showcase to draw foreign aid.

However, with strong internal pressure and external conditionalities as well as liberal policies adopted by democratically elected governments, decentralized forest policies were adopted and implemented.

Historically, there used to be, and in some cases still are, indigenous practices in which communities manage their own forest resources. However, in the Rana familial oligarchy regime, which reigned for over a century until the mid-20th century, rulers had sole ownership rights to forest resources, depriving people of their customary rights. These rulers encouraged deforestation and distributed the land to their favoured functionaries in order to generate revenue for their lavish living and to retain their power. The regimes that followed have also used forestry for their political interests, leading to massive deforestation during times of political instability, such as political movements, national referendums and general elections. This problem is very severe in the *terai*, where the government remains reluctant to implement decentralization policies. One reason the lessons of successful decentralized forestry in the hills have not been fully capitalized on in the *terai* has to do with the high-value forest resources there. Due to the lack of rights to management, local users are powerless spectators before powerful networks of illegal loggers who operate often beneath the very eyes of state authorities.

Government still holds several rights, including land tenure. Abrupt decisions on the part of government such as imposing taxes on the revenues from forest product sale often makes a mockery of the devolved system, at times breeding widespread suspicion and unpredictability.

Even where the community programme is successfully implemented, benign forest management should involve balancing human and environmental needs. So far community forestry in Nepal has been more concerned with the social aspects of communities, including poverty and livelihood issues; relatively little attention has been given to sustaining the environment and exploring the possible contributions of community forestry to this end.

Despite having some successes, many studies on community forestry in Nepal have indicated that poor people are still not benefiting equitably. In some cases, their livelihoods have even been adversely affected (Edmunds and Wollenberg, 2001; Malla, 2001; Nurse et al, 2004). Studies have shown that local elites are benefiting most because they hold the powerful positions in the executive committees and can manipulate decisions in their own favour, ignoring the agendas of the poor and marginalized. The big challenge in community forestry, then, is how to overcome elite capture and make the whole process more equitable.

Another challenge, related to elite capture, is to make collective decision-making more transparent and participatory. Executive committee members often make decisions without allowing discussion in the general assembly. This practice needs to be discouraged if the voices of women, *Dalits* and poor and disadvantaged members are to be equally taken into account.

One of the biggest challenges in recent years has been that both the government of Nepal and parallel government structures created by the Maoists have shown their distaste for forest user groups' autonomy. In particular, the financial and decision-making autonomy granted to forest user groups by the Forest Act of 1993 is being undermined by Maoists and the government.

Conclusion

In general, there are plenty of reasons to call community forestry in Nepal one of the most successful decentralized modes of forest governance, but more attention needs to be paid to making forest user groups more equitable, inclusive and pro-poor in practice. The existing policies and legislation have provided a legal grounding for decentralization of forest management roles and responsibilities from the state to the local communities, but frequent unilateral governmental policy amendments make forest user groups sceptical about their rights. Now forest user groups must work to keep their autonomy in a changing political context; they need to find a way to become politically non-threatening, but committed to democracy and protection of the rights of forest users.

Over the last two decades, the institutionalization process of forestry decentralization in Nepal has been adversely affected by unstable politics on the one hand and the techno-bureaucratic structure of forest departments on the other. In addition, the process has remained virtually dead during the last decade due to the Maoist insurgency and the resulting civil war in Nepal, which rendered the government almost non-functional at all levels. The existence of a state within a state (the Maoist people's government and the Nepal government) during the insurgency period not only created difficulties for forest user groups in making decisions at the local level about the management of forest resources, but also hindered the whole institutional process of forestry decentralization in Nepal.

Hence many challenges and limitations in community forestry in Nepal are historically rooted in inconsistent policies, unstable broader political governance and a weak institutional structure. Without understanding the context and properly addressing issues beyond forestry, those committed to decentralization of forest governance in Nepal, even in the post-war situation, will not be able to achieve their desired outcomes.

References

Agrawal, A. and Ostrom, E. (2001) 'Collective action, property rights, and decentralization in resource use in Nepal', *Politics and Society*, vol 29, no 4, pp485–514

Capistrano, D. and Colfer, C. J. P. (2005) 'Decentralization: Issues, lessons and reflections', in C. J. P. Colfer and D. Capistrano (eds) *The Politics of Decentralization: Forests, Power and People*, Earthscan, London

Department of Forests (2003) *The Community and Private Forestry Programme in Nepal*, HMG/N Publications, Kathmandu

Department of Forests (2005) *The Community and Private Forestry Programme in Nepal*, HMG/N Publications, Kathmandu

Edmunds, D. and Wollenberg, E. (2001) 'Historical perspectives on forest policy change in Asia: An introduction', *Environmental History*, vol 6, no 2, pp190–212

FECOFUN (2005) 'Study on institutional development of forest user groups in Nepal', occasional study report, Kathmandu

Fisher, B. (1999) 'Creating space: Development agencies and local institutions in natural resource management', *Forests, Trees and People*, vol 22, pp4–11

Gautam, A. P., Shivakoti, G. P. and Webb, E. L. (2004) 'A review of forest policies, institutions, and changes in the resource condition in Nepal', *International Forestry Review Journal*, vol 6, no 2, pp136–148

HMG/N (1988) 'Master plan for the forestry sector', HMG/N, Kathmandu

Hobley, M. (1996) *Participatory Forestry: The Process of Change in India and Nepal*, Rural Development Forestry Study Guide 3, ODI Publications, London

Kanel, K. R. (2006) 'Current status of community forestry in Nepal', paper submitted to the Regional Community Forestry Training Centre for Asia and the Pacific, Bangkok

Malla, Y. B. (2001) 'Changing policies and the persistence of patron-client relations in Nepal', *Environmental History*, vol 6, no 2, pp287–307

Malla, Y. B., Neupane, H. R. and Branney, P. J. (2003) 'Why are not poor people benefiting more from community forestry?', *Journal of Forests and Livelihoods*, vol 3, no 1, pp78–90

MoFSC (1995) 'Forest Act (1993)' and 'Forest Regulation (1995)', Ministry of Forest and Soil Conservation, HMG/N Publications, Kathmandu

Nurse, M., Robinson, P., Paudel, D. and Pokharel, B. K. (2004) 'Towards pro-poor community forestry – Recent experiences from Dolakha and Okhaldhunga Districts', 'Proceedings of the National Workshop on Management of Common Property Resources and Equity: Exploring lessons from Nepal, 28 May 2003', occasional paper, Forest Action, Kathmandu, Nepal, pp41–58

Pokharel, B. K. and Nurse, M. (2004) 'Forests and people's livelihoods: Benefiting the poor from community forestry', *Journal of Forests and Livelihoods*, vol 4, no 1, pp19–29

Ribot, J. C. (2004) *Waiting for Democracy: The Politics of Choice in Natural Resource Decentralization*, World Resources Institute, Washington, DC

Springate-Baginski, O., Yadav, N. P., Dev, O. P. and Soussan, J. (2003) 'Community forest management in the middle hills of Nepal: The changing context', *Forest and Livelihood*, vol 3, no 1, pp5–20

6

Policies and Programmes to Make Decentralization Effective: A Case Study from China

Liu Dachang

Decentralization of natural resource management varies from one country to another in terms of transfer of natural resources, responsibility and power: from central government to local governments, and from government to community-based organizations, individual households or the private sector. In China, decentralization in the forest sector has taken place mainly in collective forests, the largest category of forest ownership from the mid-1950s to the early 1980s. The first decentralization took place in the early 1980s (Liu, 2001; Liu and Edmunds, 2003a and 2003b). Drawing on the experience of the past two decades, China launched a new round of decentralization in collective forest resources in the early part of this century. This decentralization process is expected to be complete by 2010.

Many have examined the experience of decentralization in the 1980s to understand its consequences and explain why it failed to achieve the results policymakers originally desired (Menzies and Peluso, 1991; Sun, 1992; Bruce et al, 1995; Song et al, 1997; Yin and Newman, 1997; Albers et al, 1998; Zhang et al, 2000; Liu and Edmunds, 2003a and 2003b). Some of the lessons learned from the experience have been incorporated into new policies for the ongoing decentralization.

However, decentralization cannot be understood in isolation from the government's other policies and programmes, even within the forest sector. Although decentralization in collective forests is intended to increase household forest tenure security and power, other forest policies and programmes may undermine the effort. In particular, policies on the provision of ecological services, conservation of nature and biodiversity, and control over forest exploitation have different objectives and often directly contradict household tenure and power.

This chapter examines whether decentralization is compatible with the policy for public benefit forests (*shengtai gongyilin*), the policy for nature reserve establishment, the logging permit policy and the Natural Forest Protection Programme. It then analyses their effects on the success of decentralization and demonstrates the need for better coordinated forest policies and programmes. The remainder of this chapter provides a brief review of decentralization in the early 1980s and today, examines inconsistencies between other policies and programmes and decentralization, and discusses potential options for coordinated forest policies and programmes to balance competing management objectives. The final section makes some suggestions for the future.

DECENTRALIZATION IN COLLECTIVE FORESTS

There were only two types of forest ownership in China from the mid-1950s to the early 1980s: state forest and collective forest. State forests made up about 45 per cent of China's total forest area at that time, and collective forests about 55 per cent (Table 6.1).

First decentralization

The decentralization in the early 1980s was a limited individualization of collective forests. It involved a transfer of forest land which had no forest cover and responsibility for and limited power over existing forest management. In the government-driven decentralization process, permanent use rights to collective non-forested lands, or open forest lands and shrubland where a village had no or little non-forested lands, were distributed to farm households for them to plant trees to meet their needs for fuelwood and timber. Forest lands so distributed are called *ziliushan* ('family plots'). Meanwhile, parts of existing collective forests were allocated to farm households (and groups of households in some cases) as an effort to improve forest management. The forests so allocated are *zerenshan* (translated as 'responsibility hills' in this author's previous papers). As a result of the transfer and later developments, China now has three types of forest ownership: state forest, collective forest, and household and private forests (Table 6.1).

Table 6.1 *China's forests, by ownership*

Mid-1950s to early 1980s		*2006*	
State forests	45%	State forests	42%
Collective forests	55%	Collective forests	38%
		Household and private forests	20%
Total	100%	Total	100%

Sources: Ministry of Forestry (1990) and SFA (2005).

The decentralization transferred limited power over forest management and marketing to farm households. The changes in tenure, power and roles of stakeholders are summarized in Table 6.2 (see also Liu, 2001; Liu and Edmunds, 2003a).

Second decentralization

A new round of decentralization was launched after the central authorities issued 'Decisions regarding speeding up forestry development' (Central Committee of CPC and the State Council, 2003). A few provinces, such as Fujian, Jiangxi and Liaoning, are far along in the decentralization process, but most others are still in a pilot phase.

Box 6.1 Policy on reform of collective forest management in Jiangxi Province

Farm households possess permanent and inheritable use rights to family plots. If the village collective regained them after their initial distribution and farmers want to resume their care, return the family plots to households. If they were combined with *zerenshan* forests and farmers want to split them, let them do so. If no family plots were distributed in the forest reform in the 1980s and if farm households now want family plots and there are forest lands available for distribution, distribute family plots to them.

Return *zerenshan* forests to farm households if the village collectives regained them after their initial distribution and farmers want them back. Farm households own the trees in the forests and this ownership right is inheritable during the term (30–70 years).

Return family plots and/or *zerenshan* forests to farm households once existing trees or plantations on them are harvested if the village collective regained them because of households' failure to plant trees on them.

Distribute remaining collective forests equally to farm households (or groups of farm households) as *zerenshan* forests. Or distribute them equally, in the form of a monetary share, to all residents in the village, if collective management is to be maintained and farmers are satisfied with and agree to the tenure and management arrangement. Then distribute income from the forests to households based on their shares.

Management of collective forests can be contracted out to individuals or other entities, internal or external. In that case, the village collective will collect contract fees or income annually from the contractor and distribute 70 per cent or more of the income equally to farm households as owners and use the remaining amount for village public welfare projects and other expenses.

Source: Provincial Government of Jiangxi (2004)

Table 6.2 *Forest tenure and stakeholders' rights under decentralization*

Tenure types	*Stakeholders*	*First decentralization (early 1980s)*	*Second decentralization (2004–2010)*
Ziliushan ('family plot' forest)	Village collective	Landowner	Landowner Use rights to family plots returned to farm households (see Box 6.1)
	Households	Held permanent, inheritable land-use rights Owned the trees planted on the land Made decisions about trees to be planted and about harvest (after securing logging permit) Could not transfer land-use rights	Can sell or lease land-use rights and timber and use forest as mortgage or collateral
Zerenshan ('responsibility hill' forest)	Village collective	Owner of forest (including land and trees)	Forest landowner only
	Households	Managed forests allocated to them for an unspecified period of time and shared forest products and incomes with village collective Had limited control over harvest and sale of trees Could not transfer the forests	Hold land-use rights for 30–70 years and own trees on their forests Manage and gain benefits from their forests Or collectively own forest but contract its management to other entities and share income equally among them Can sell and lease land-use rights and timber and use them as mortgage or collateral

Sources: Liu, 2001; Liu and Edmunds, 2003a and 2003b

The ongoing decentralization aims for a greater individualization and private ownership of collective forests. Forest tenure is at its core, with forests transferred to individual farm households, in contrast to the transfer of non-forested lands only during the first decentralization in the early 1980s. Land-use rights to *zerenshan* forests are distributed equally (in terms of area per person) to farm households for a period of 30 to 70 years, and trees on the land are also household property. This makes tenure of *zerenshan* forests almost the same as that of family plots. The only difference between the two now is that farmers hold land-use rights to family plots permanently, whereas they must renew land-use rights to *zerenshan* forests by contract with the village collectives when the present contracts expire.

Another feature of decentralization is the adoption of multiple forest tenure and management arrangements in a province to allow farmers to choose an arrangement suitable for local socioeconomic circumstances. Thus a shareholding system (Liu and Edmunds, 2003b) is still recognized as a tenure and management arrangement, if households prefer it. If this is the case, farm households own forest in the form of a share. For instance, if a village consists of 50 households and it divides its forest into 1000 shares, then each household will hold 20 shares. Nevertheless, the forests are still managed collectively. Accordingly, the households, as shareholders, obtain income from the forests based on their shares. In addition, other tenure and management arrangements are allowed; but whatever arrangements are adopted, forest incomes must be distributed equally to farm households as forest owners (see Table 6.1 and Box 6.2 for more information on forest tenure and management arrangements).

The current decentralization gives farmers greater power to make decisions about management of their forests (planting and tending, for example). They are also allowed to sell and lease their land-use rights, sell their standing trees and use these forest properties as investment (shares) in forestry and other undertakings and as collateral to obtain loans from banks (Table 6.2; Box 6.1). Although these powers were an innovation in the early 1990s and legalized by the revised Forest Law in 1998, the government is now developing transaction markets for forest properties. It is too early to assess the impact of these policy changes, but reports of increasing inputs of labour and funds for tree planting and forest management and growing numbers of transactions indicate a positive response to the policy changes. For example, in Yong'an, Fujian Province, transactions involving 14,000ha of forests worth 131 million Yuan (US$16.380 million) were made in a short period from October 2004 to June 2005, and loans of 120 million Yuan (US$15 million) were granted against collateral of trees and land-use rights (*China Green Times*, 5 July 2005, pA2). Moreover, farmers are now allowed to sell their wood and bamboo timber in free markets, whereas in the past the state had a monopoly on purchases.

Other forest reforms complement the current decentralization in collective forests. One matching reform is a reduction in forest taxes and fees to ensure that farmers as forest owners gain a fair share of income from their forest property, especially wood. Besides the agriculture tax, China has levied a 'special agricultural products tax' since 1994 to tax forest products. In addition, the government charged more than 10 other official types of taxes and fees on timber and wood commodities and several unofficial forestry charges (Liu et al, 2003; Liu and Landell-Mills, 2003). As a result, in the case of wood, the government collected 50 per cent or more of the price; farmers received only about 20 per cent for their care for the forests, with the remaining 30 per cent covering the costs of logging and wood transport. The high taxes were a serious disincentive for tree planting and forest management and had negative social and environmental effects (SFA Survey Team, 1999; Zhang, 1999; Liu and Edmunds, 2003a; Liu et al, 2003; Liu and Landell-Mills, 2003).

As an integral part of the agriculture tax reform, the government has now abolished the special agricultural products tax and all unofficial types of fees to increase farmers' share of forest income and hence improve welfare and create incentives for tree planting and forest management (Box 6.2). Hence the new policy for income distribution is often – and rightly – dubbed 'the farmers' right to obtain benefits from forests'. If it is fully implemented, the policy is likely to encourage farmers to invest more in plantation establishment and forest management and increase the return on their investment in their forests.

Another necessary reform encompassed by the current decentralization process is improvement in the forest services (Central Committee of CPC and the State Council, 2003; Provincial Government of Jiangxi, 2004). Indeed, improvements

Box 6.2 Policy on reform of collective forest management in Jiangxi Province

Abolish the special agricultural products tax. (The tax was abolished throughout China from 2006 after a few years of a pilot phase.) Cancel all kinds of fees on wood and bamboo levied by prefecture, county and township governments and village administration.

Cut the Reforestation Fund (a fund established by the central government for investment in reforestation, collected from farmers and other timber producers) by setting a lower base rate on which the fund is collected. (The base rate on which the fund is collected from commercial timber is now set at 360 Yuan (US$45) per m^3, compared with about 500 Yuan (US$65) per m^3 previously.)

Exempt value-added tax on round wood and bamboo if they are for on-farm use (not for sale in markets). Income tax is also temporarily exempted (for an unspecified period of time).

Source: Provincial Government of Jiangxi (2004)

are needed in the areas of technical services and forest extension, policy advice, and market and financial information. More needs to be done to understand how the forestry agencies can better serve farmer-foresters.

Why Decentralization Now?

An important argument for the current decentralization relates to the historical exclusion of farm households from collective forest management, as noted by Edmunds et al (2003), but it has little to do with other reasons such as cost reduction (Colfer, 2005) or donor pressure (Agrawal, 2002). Decentralization has several specific objectives:

- increasing forest growth (both area and stocking levels) and encouraging sustainable forest management;
- promoting rural development and job creation for farmers through higher forest productivity; and
- improving farmers' livelihoods and thus reducing poverty.

Policymakers and advocates of decentralization note that China has been able to produce enough food for its huge population from its 120 million hectares of agricultural land but has failed to provide an adequate level of forest products and ecological services from its 280 million hectares of forest land – more than double the extent of agricultural land. They want China's forest sector to provide more forest products and contribute more to rural livelihoods, regional economic development and ecological services.

Conflicting Policies and Programmes

Public benefit forests

China has adopted a new strategy for forest management: to manage forests based on their primary function (see Table 6.3). In line with the strategy, forests in China are classified into two broad categories – production forests and forests for public ecological benefits – each with distinct management objectives. Production forests are managed mainly for timber, fuelwood and other forest products. Public ecological benefit forests are managed primarily for ecological services; hence they are strictly protected, and logging and collection of other forest products is heavily regulated. Central and local governments need to compensate owners of public benefit forests for their loss of opportunities for harvest of forest products; in other words, the governments (on behalf of the public) must pay owners for the ecological services their forests provide.

Table 6.3 *China's forests, by primary function*

Production forests	Timber forests Fuelwood forests Economic forests (non-timber forest products) Bamboo forests
Public benefit forests	Protection forests Special purpose forests

Note: Besides forests, there are other types of forest land: open forest land, shrubland, young plantation (unestablished forest), nursery land and non-forested land. Production forests are from young plantations (unestablished forests) and can be developed on open forest land and non-forested land.

The ideas that forests should be managed according to their primary functions and that payments should be made for environmental benefits were first put forward in China in the 1980s and were officially recognized in the early 1990s (State Council, 1993) and incorporated into the revised Forest Law of China in 1998. Implementation of the compensation policy started in 2001 in 658 counties in 11 provinces and 28 national nature reserves, with the central government allocating 1 billion Yuan (US$125 million) for environment payments in that year. Since then, more forests have been classified as public benefit forests. The central government has increased funds for such forests, and local governments are required to allocate funds to designate additional public benefit forests. Forests covered by central government funds are called key public benefit forests; those covered by local governments are called local public benefit forests. Most provinces are in the process of demarcating public benefit forests and production forests and have also allocated limited funds for payment for local public benefit forests.

As a consequence, many collective forests and household forests are being designated as public benefit forests. For example, Guizhou Province has designated 2.0 million hectares of public benefit forest but has only about 0.4 million hectares of state forests (10 per cent of the total forest area of 4.2 million hectares). Thus, even if the entire 0.4 million hectares of state forest is managed as public benefit forest, the remaining public benefit forest (1.6 million hectares) will be created from collective and household forests. Similarly, Zhejiang Province has also designated 2.0 million hectares of public benefit forest but has only about 0.5 million hectares of state forests. Liaoning Province has designated 3.5 million hectares of public benefit forest but has only about 0.5 million hectares of state forests.

In the ongoing decentralization process, many provinces distribute only those collective forests that are defined as production forests to farm households, and a few provinces distribute both production forests and public benefit forests. Each approach has implications for the individualization of collective forest. When only production forests are distributed, limited or even no forest is available for distribution to farm households, since a large proportion of collective forests is designated as public benefit forest. When both production forests and public benefit forests are distributed, farm households own and manage designated public benefit forests but have no rights to logging and collection of other products. Although they are compensated for the loss of harvesting opportunities, the compensation is minimal. At present, the central government compensation stands at 75 Yuan (less than US$10) per hectare per year, and local government compensation for local public benefit forests is even lower. The central government compensation covers only part of the costs of forest management, such as tending and forest patrol, and does not equal the market value for the ecological services that public benefit forests provide. It is said that farm households should share the costs of ecological services because they also enjoy the ecological services. But as it stands now, they cover the overwhelming majority of costs. The high opportunity costs for owners will certainly be a disincentive for sustainable forest management and will fail to help owners generate income and improve livelihoods – one objective of decentralization. In other words, ecological services provided by public benefit forests should be marketable and paid for at a reasonable rate of compensation if decentralization is to achieve its objectives.

Expansion of protected areas

To date, there has been only one type of protected area in China: nature reserves. The first nature reserve in modern-day China was established in Guangdong Province in 1956. As shown in Table 6.4, the designation of new nature reserves was very slow before the late 1980s, and the area of nature reserves accounted for less than 3 per cent of China's total land area in 1989. Since then, however, the designation of nature reserves has accelerated significantly, especially from 1999 to 2000. At the end of 1999, there were 1119 nature reserves accounting for 8.8 per cent of China's total land area, compared with 763 nature reserves accounting for 6.8 per cent of total land area in 1993. The year 2000 saw a remarkable expansion, with an increase of more than 3.5 per cent in one year. The year 2001 also saw a growth of nearly 2 per cent in the number of nature reserves.

The trend of nature reserve designation for Yunnan, a mountainous province of rich biodiversity, is similar. Yunnan established its first nature reserve in 1958. In 2000, nature reserves in Yunnan totalled 114, covering about 6 per cent of the total land area. Among the 114 nature reserves, 109 are terrestrial forest. By the end of 2005, the number of nature reserves rose to 198, encompassing 3.55 million hectares and accounting for 9 per cent of the province's land area, with an overwhelming majority of the nature reserves still forests.

The expansion occurred in two forms: an increase in the number of nature reserves and an expansion of the area of existing nature reserves. For example, over a period of less than two years, during 2004–2005, 24 new nature reserves were

Table 6.4 *Nature reserves in China, 1956–2000*

Year	Number	Area (million ha)	Percentage of total land area
1956	1	0.001	
1965	19	0.649	0.07
1978	34	1.265	0.13
1982	119	4.082	0.43
1989	573	24.763	2.58
1993	763	66.184	6.80
1997	926	76.980	7.64
1999	1119	88.100	8.80
2000	1276	123.000	12.44
2001	1511	144.700*	14.40*
2005	2349	143.949	14.99

Sources: Department for Plant and Wildlife Conservation, SFA (2003); Department for Nature and Ecology Conservation, State Environment Protection Administration (2005); *China Green Times*, 1 September 2001, pA1, and 13 September 2002, pA1.
*Including area of marine nature reserves.

designated in Xiangxi Autonomous Prefecture, Hunan Province. The number of nature reserves rose to 46 in total, covering about 20 per cent of total land area of the prefecture (*China Green Times*, 15 July 2005, pA2). When Zhujiangyuan ('Source of the Pearl River') Nature Reserve in Yunnan was upgraded from a county-level nature reserve to a provincial nature reserve in 2000, its area increased 100-fold.

Given the long-standing and extensive human habitation in mountainous areas, the establishment of protected areas has often meant enclosing part or all of certain communities within nature reserves. Nationally, 1699 nature reserves were established and managed by forestry departments at central and local levels by the end of 2005, with a total area of 120 million hectares (*China Green Times*, 4 September 2006, pA2). By way of reference, China's forest area totals 169 million hectares, of which 74.34 million hectares is state forests and 99.94 million hectares is collective and household forests (SFA, 2005). The spread of nature reserves has enclosed many collective forests and household forests inside nature reserves (about 45 million hectares), as these examples show:

- Sanjiangyuan ('Source of the Three Rivers') Nature Reserve covers half of the total area of Qinghai Province in west China, but since it is impossible that all the land is state owned, some proportion is collective.
- 144 nature villages are enclosed in Wuliangshan Nature Reserve, a national nature reserve located in Yunnan. A survey found that about 330ha of forests belonging to 18 sample nature villages were enclosed in the nature reserve (FCCDP, 1999a).

- A survey found that 3300ha, or 51 per cent, of collective forests of nine sample nature villages in Nuozhadu, another national nature reserve located in Yunnan, was enclosed in the nature reserve (FCCDP, 1999b).
- In 2005 two villages, comprising 132 households, were completely within the core and buffer zones of the national Huapin Nature Reserve in Guangxi (*China Green Times*, 23 May 2005, pA1).

This tenure problem with nature reserves largely relates to the fact that the majority of nature reserves were established after the first decentralization (distribution of collective forests to households) in the early 1980s. The State Forestry Administration's Department for Plant and Wildlife Conservation (2003) agrees that the tenure problem is a great challenge for effective nature reserve management.

Nature reserves are strictly protected through laws and regulations. Logging and harvest of forest products, including non-timber forest products, are prohibited. The proliferation and expansion of nature reserves deprive collective and household forest owners of their legal property and use rights (Box 6.3), despite the statement that ownership and use rights would not change (Ministry of Land Resource Management, 1995). Clearly, the management objectives of nature reserves – nature and biodiversity conservation – are not compatible with decentralization aims, which are to strengthen the property rights of rural households.

Box 6.3 Uses of collective forest and household forest in Wuliangshan Nature Reserve

Liangjiadi is one of 144 nature villages enclosed in the nationally designated Wuliangshan Nature Reserve, located in southwestern China's Yunnan Province. The village has about 27ha of forest, half of which was distributed to households as family plots in the early 1980s. But later, all of the forest area was enclosed in the nature reserve, and logging from it is no longer allowed. As a result, villagers must now purchase timber for on-farm use from other villages.

Source: FCCDP (1999a)

Villagers in remote mountainous areas, often designated as nature reserves, are farmers who rely heavily on forests for livelihoods, including fuelwood for cooking and heating and other essentials. Denying villagers access to their forests leaves them with fewer opportunities for employment and income generation, and if they harvest their forests, they risk being prosecuted. The conflicting management objectives often lead to conflicts between the communities and nature reserve advocates and management agencies.

Moreover, increasing numbers of wild animals protected in the expanded nature reserves damage villagers' crops and livestock and severely injure or even kill villagers. But public compensation for such damage is quite limited or non-existent.

It is necessary to assess whether distribution of collective forests within nature reserves to households is appropriate. If not, other options that strengthen the forest property rights of households and communities should be sought, including full exercise of their rights to the family plots distributed to them in the early 1980s. Options for addressing the issue could include the following:

- raising the government's compensation for public benefit forests to a reasonable rate and paying it directly to villagers (rather than to the nature reserve agency) for loss of opportunities for product harvesting;
- allowing villagers to participate as paid labourers in various nature reserve management activities;
- identifying and developing alternative economic opportunities for villagers; and
- relocating villager households on a voluntary basis to areas outside the nature reserves.

The Natural Forest Protection Programme

China officially launched the Natural Forest Protection Programme in October 2000 after a two-year pilot phase (from October 1998 to September 2000). The programme was developed in response to the disrupted flow of the Yellow River in 1997 and severe floods on the Yangtze River in 1998. It is a 10-year programme, implemented from 2001 to 2010, involving a planned investment of 96.2 billion Yuan (US$12 billion), or more than 100 billion Yuan if the costs of the pilot phase are included (State Council, 2000a and 2000b). The amount is even greater if bad loans to state forest corporations, cancelled as a result of the logging ban, are taken into account.

The primary objectives of the Natural Forest Protection Programme are reforestation and environmental restoration of existing forests – goals that conflict with those of the current decentralization, which seeks to increase household forest tenure security and give households more power over forests. Often misinterpreted as a logging ban, the programme actually has three components:

1 a ban on all logging in the upper Yangtze River (Chongqing, Guizhou, Hubei, Sichuan, Tibet and Yunnan), mid to upper Yellow River (Gansu, Henan, Inner Mongolia, Ningxia, Qinghai, Shaanxi and Shanxi), Hainan and Xinjiang and diminished logging in state-owned forests in northeastern China;
2 reforestation through large-scale planting of trees (including sowing by aeroplane) and natural regeneration of secondary forests through mountain

closure (by prohibiting or restricting grazing and harvesting on degraded mountainous land) as well as forest tending in the programme area; and

3 provision of alternative employment (from forest patrol to tree planting and tending) and pensions for state logging corporations that implement the programme.

The programme has achieved its objectives of ecological restoration in the first phase (2001–2005). The trend in forest depletion was effectively reversed and the forest area and standing stock increased. For instance, in Yunnan, commercial wood harvest decreased to virtually zero in 2000 from 4.04 million cubic metres in 1997. The programme, together with other forest programmes and efforts, also contributed to an increase of about 5 per cent in forest cover in Yunnan from 2001 to 2005. Employees of the state logging corporations now play the role of tree planters and forest caretakers or have been transferred to other sectors.

The logging ban has been applied not only to state forests but also to collective and household forests. Though their rights to forest are recognized, farmers were not consulted about whether their forests should be included in the programme; instead, the governments defined the programme area. Moreover, the programme's title clearly indicates that it is a ban on commercial logging from natural forests, but in practice it was extended to planted forests, including a ban on harvest for non-commercial use and tight control over fuelwood harvest. Farmers are not allowed to harvest and sell wood from their forests. As a consequence, they have experienced a significant decline in income and employment opportunities, and many farm households are falling below the poverty line. The first years of the programme also saw increased short-, mid- and long-term social costs, including a less-skilled workforce because of school dropouts, for instance. Nevertheless, all the programme funding went to the state forestry corporations that implemented it – and none to the affected farmers.

Like the nature reserve policy discussed above, the Natural Forest Protection Programme also raises the question of communities' and rural households' legal property rights: do they have the right to dispose of their forest property as they wish? Farmers should be able to exercise their legal rights to their trees as much as possible, by, for example, receiving priority when the ban is gradually lifted (currently in the pilot phase) and logging quotas are distributed. Strengthening forest ownership of farm households alone is not enough to encourage them to invest in tree planting and sustainable forest management.

The Natural Forest Protection Programme also prevents full implementation of decentralization. It covers 17 provinces making up 81 per cent of China's land area. Yunnan Province has already announced that only planted production forests will be distributed to households in the programme area and that collective natural forests cannot be distributed, which means collective forests available for distribution are quite limited. It is unclear what policy other provinces will make in relation to decentralization in collective forests in the programme area, but very likely they will do the same as Yunnan. One solution is to transform this long-term programme by incorporating it into the policy for public benefit forests and paying proper compensation for ecological services directly to village communities and households rather than to the state forest corporations, as currently practised.

Attention must also be given to farmers' participation in programme planning rather than simply including their forest land in the programme – consistent with the goal of decentralization to give more power over forests to rural households.

Log harvest quotas

A log harvest quota was first introduced in 1985 and has been fully implemented since 1987 to control timber harvesting (Ministry of Forestry, 1987). The policy was incorporated into the revised Forestry Law of China in 1998.

The basic principle for the quota policy is that annual wood consumption must be less than annual forest growth. An annual quota is set once every five years, and the quota plan for 2006–2010 is now being implemented. The county is the basic level at which the quota is set, and its plan is then sent to the provincial departments of forestry, which in turn set provincial quotas. The provincial plans are then sent to the State Forestry Administration, which sets the national quota. The county and province plans are subject to significant adjustment so that the total harvest does not exceed the national quota. The national quota must be endorsed by the central government (see Liu, 2001, for details on quota planning).

The scheme sets annual harvest volume, in total and by type. The annual harvest quota comprises five types: commercial timber, timber for on-farm use, wood for sideline production such as mushroom cultivation, fuelwood and wood for other uses. Volumes are not allowed to exceed the quota, and, in general, redistribution of quotas between types of wood consumption and redistribution of quotas between years are not allowed, though exceptions do exist (Ministry of Forestry, 1990; State Council, 2006).

Although China is not unique in having forest harvest quotas, it is unique in applying the quota scheme to forests of all ownership types, including collective forests and household forests. The quota is distributed by consumption type, not by forest ownership. Owners and managers of collective and household forests have greater difficulty gaining access to the quota than state corporations, since they have less connection with authorities and less information.

Implementation of the quota policy raises another concern. The centrally approved quota is allocated back down through county bureaus of forestry to the township level, which in turn distributes its quota among its villages. But the entire harvest quota for a township is often allocated solely to forest farms and traders, which means that individual households (as forest owners and managers) are not able to obtain permission to harvest their forests. It is also very difficult for owners and managers of collective forests and especially household forests to obtain logging permits. The many required documents on harvest planning and reforestation planning and the time-consuming travel necessary to apply put logging permits out of reach for many.

The logging permit policy and its implementation process diminish households' forest tenure security and management incentives and are thus not consistent with the stated objectives of decentralization. The logging permit policy and implementation process discourage owners and managers of collective and household forests from establishing timber plantations. In response, they have chosen to develop economic forests (for non-timber forest products) that are subject to fewer regulations. For instance, when the government launched the nationwide Programme for Conversion of Steep Farmland to Forest in the late 1990s, farmers gave priority to the development of economic forests, although forestry departments were seeking a balance between economic forests and forests for wood and environmental services. In other words, the logging permit policy together with the implementation process is making decentralized forest management less effective.

The current decentralization process is moving towards a transparent process and easier procedures for forest farmers when harvest quotas are distributed. If this continues, owners and managers of collective forests and household forests may have better access to the quota. Classifying quotas by forest ownership would be another option to strengthen household forest tenure security and encourage household incentives for forest development – making decentralization more effective.

Conclusion

China's experience in decentralization shows that to achieve the objectives, forest policies and programmes need to be coordinated and compatible. Decentralization through privatization is but one of the forest policies contributing to sustainable forest management for environmental services and livelihood improvement. If forest policies and programmes are not compatible, none of them – including the decentralization policy – will be effective. Clearly, this calls for defining the legal basis of property rights for private and quasi-private forest management to supply ecological and other services from forests, and developing consistent laws and policies for nature and biodiversity conservation. Further study is needed to determine whether decentralization through privatization is appropriate for biodiversity conservation, whether ecological services provided by villagers' forests should be marketable and paid for, and whether China needs to seek additional harmonized approaches.

The Chinese experience over the past two decades also indicates that privatization may be more efficient than the previous system in producing commercial forest goods, and may be more conducive to private investment in the sector. But this will be the case only if forest owners have the right to dispose of their forest property through sale, lease, collateral for loans and shares as investments. A conducive environment for investment in forestry also includes distribution arrangements of forest income that are fair to farmers. The policies for harvest and marketing of forest products must be consistent with the secure household forest tenure that decentralization intends to achieve. Farmers should be given more power over forests; ownership alone is not enough.

References

Agrawal, A. (2002) 'The regulatory community: Decentralization and the environment in the *van panchayats* (forest councils) of Kumaon', paper presented at the World Resources Institute Workshop on Decentralization and the Environment, 18–22 February, Bellagio, Italy

Albers, H. J., Rozelle, S. D. and Li, G. (1998) 'China's forests under economic reform: Timber supplies, environmental protection, and rural resource access', *Contemporary Economic Policy*, vol 16, pp22–23

Bruce, J. W., Rudrappa, S. and Li, Z. (1995) 'Experimenting with approaches to common property forestry in China', *Unasylva*, vol 46, no 180, pp44–48

Central Committee of CPC and the State Council (2003) 'Decisions regarding speeding up forestry development', Beijing

China Green Times, 1 September 2001, pA1; 13 September 2002, pA1; 23 May 2005, pA1; 5 July 2005, pA2; 15 July 2005, pA2; 4 September 2006, pA2

Colfer, C. J. P. (2005) *The Complex Forest: Communities, Uncertainty and Adaptive Collaborative Management*, Resources for the Future, Washington, DC, and Center for International Forestry Research, Bogor, Indonesia

Department for Nature and Ecology Conservation (State Environment Protection Administration) (2005) *China Nature Reserves*, China Environmental Science Press, Beijing (in Chinese)

Department for Plant and Wildlife Conservation (State Forestry Administration) (2003) *Study on China's Nature Reserve Policies*, China Forestry Press, Beijing

Edmunds, D., Wollenberg, E., Contreras, A. P., Liu, D., Kelkar, G., Nathan, D., Sarin, M. and Singh, N. M. (2003) 'Introduction', in D. Edmunds and E. Wollenberg (eds) *Local Forest Management: The Impacts of Devolution Policies*, Earthscan, London

FCCDP (1999a) *Field Survey Report: Wuliangshan Nature Reserve*, FCCDP, Kunming, China (in Chinese)

FCCDP (1999b) *Field Survey Report: Nuozhadu Nature Reserve*, FCCDP, Kunming, China (in Chinese)

Liu, D. (2001) 'Tenure and management of non-state forests in China since 1950: A historical review', *Environmental History*, vol 6, no 2, pp239–263

Liu, D. and Edmunds, D. (2003a) 'Devolution as a means of expanding local forest management in south China: Lessons from the past 20 years', in W. F. Hyde, B. Belcher and X. Jiantao (eds) *China's Forests: Global Lessons from Market Reforms*, Resources for the Future/CIFOR, Washington, DC/Bogor, Indonesia

Liu, D. and Edmunds, D. (2003b) 'The promises and limitations of devolution and local forest management in China', in D. Edmunds and E. Wollenberg (eds) *Local Forest Management: The Impacts of Devolution Policies*, Earthscan, London

Liu, D., Weng, Q., Zeng, J., Zheng, H. and Zhou, Z. (2003) 'Degradation and rehabilitation of evergreen broadleaf forest in Cangwu County, Guangxi', in Liu Dachang (ed) *Rehabilitation of Degraded Forests to Improve Livelihoods of Poor Farmers in South China*, CIFOR, Bogor, Indonesia

Liu, J. and Landell-Mills, N. (2003) 'Taxes and fees in the southern collective forest region', in W. F. Hyde, B. Belcher and X. Jiantao (eds) *China's Forests: Global Lessons from Market Reforms*, Resources for the Future/CIFOR, Washington, DC/Bogor, Indonesia

Menzies, N. and Peluso, N. (1991) 'Rights of access to upland forest resources in southwest China', *Journal of World Forest Resources Management*, vol 6, no 2, pp1–20

Ministry of Forestry (1987) *China Forestry Yearbook*, China Forestry Press, Beijing (in Chinese)

Ministry of Forestry (1990) *China Forestry Yearbook*, China Forestry Press, Beijing (in Chinese)

Ministry of Land Resource Management (1995) 'Regulations (*guanli banfa*) on land management of nature reserves', Beijing (in Chinese)

Provincial Government of Jiangxi (2004) 'Policy on reform of collective forest management in Jiangxi Province', www.jx.gov.cn, accessed May 2006 (in Chinese)

SFA (State Forestry Administration) (2005) 'China forest resources', Beijing

SFA Survey Team (1999) 'A study on timber taxes and fees: Hunan case', *Forestry Economics* (China), vol 117, pp26–35 (in Chinese)

Song, Y., Burch Jr, W., Geballe, G. and Geng, L. (1997) 'New organizational strategies for managing the forests of southeast China: The shareholding integrated forest tenure (SHIFT) system', *Forest Ecology and Management*, vol 9, no 1, pp183–194

State Council (1993) 'Further strengthen work on plantation establishment and greening', Beijing (in Chinese)

State Council (2000a) 'Implementation plan for Natural Forest Protection Program in Upper Yangtze River and Mid-to-Upper Yellow River', Beijing (in Chinese)

State Council (2000b) 'Implementation plan for Natural Forest Protection Program in key state forest area in Inner Mongolia, northeast China', Beijing (in Chinese)

State Council (2006) 'Approval of national logging quota for 2006–2010', www.gov.cn, accessed May 2006 (in Chinese)

Sun, C. (1992) 'Community forestry in south China', *Journal of Forestry*, vol 90, no 6, pp35–40

Yin, R. and Newman, D. (1997) 'Impacts of rural reforms: The case of the Chinese forest sector', *Environment and Development Economics*, vol 2, no 3, pp289–303

Zhang, X. (1999) 'The forestry taxation and charges issue in the collective forestry area of Fujian and Jiangxi provinces', *Forestry Economics* (China) vol 6, pp36–45 (in Chinese)

Zhang, Y., Kuuluvainen, J. and Uusivouri, J. (2000) 'Impacts of economic reforms on rural forestry in China', *Forestry Policy and Economics*, vol 1, pp27–40

7

Forest Fire Management and Autonomy in the Republic of Korea

Byoung Il Yoo

In the Republic of Korea (the official name of South Korea), decentralization has had significant effects on tree volume, land use and, most significantly, forest fires. Massive forest fires have caused considerable environmental damage and adversely affected the social, economic and cultural aspects of forest sustainability at both local and national levels. The central government has now strengthened arson laws and fire preparedness, prevention and suppression, and integrated incident command guidelines have been developed to ensure cooperation between the central and local governments. The central government has taken more responsibility for natural resource conservation, but in the future, systematic cooperation among the central government, local governments and communities will be required to improve natural resource management.

Administrative History

The Korean Peninsula (Figure 7.1) was divided after World War II, and in 1948 the Republic of Korea was established as a democratic country. Local autonomy and decentralization began in 1951, but after the Military Revolution in 1961, the Local Autonomy Law was temporarily withdrawn in response to social chaos. It was reintroduced in 1992 because of calls for democracy, and after several years of preparation, local autonomy was reinstated in 1995.

Local governments, according to Article 117 of the Constitution, 'shall deal with matters pertaining to the wellbeing of local residents, shall manage properties, and may establish their own rules and regulations regarding local autonomy as delegated by national laws and decrees'. Here, 'local government' refers broadly

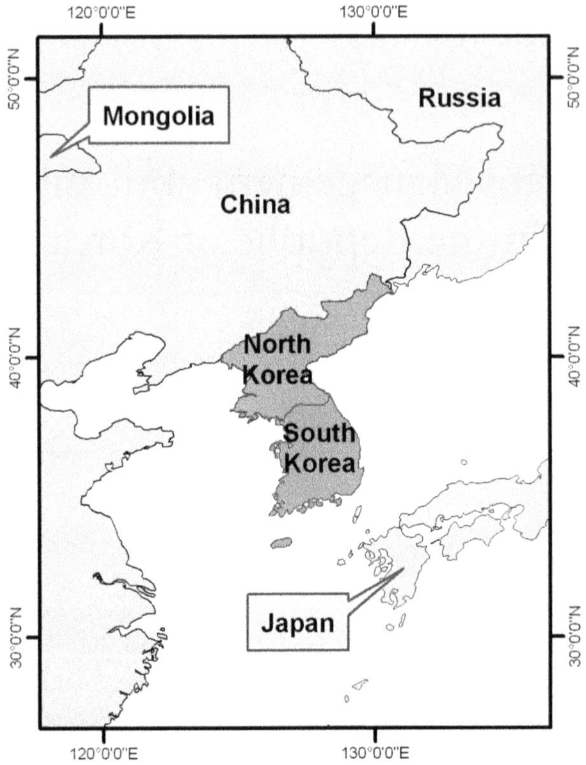

Figure 7.1 *The Korean Peninsula in northeast Asia*

Source: KFRI (2003)

to administrations at two levels, the province and the district. Provincial-level governments include one 'special' city, six metropolitan cities and nine provinces. Districts are divided into either *Si* or *Gun*, depending on size and population, which in turn are divided into the lowest-level administrative units, *Eup* and *Myun*. District-level governments include 74 *Si*, 89 *Gun*, 205 *Eup* and 1217 *Myun* (MOGAHA, 2006).

After the introduction of autonomy in 1995, the Republic of Korea planned to devolve power from the central government in Seoul (the capital city) to local governments. The regions would have autonomy in economic power, administration, finance, education, culture and resources. The process has been slow for political reasons and conflicts of interest among stakeholders. A monetary crisis, which started just after local autonomy, caused considerable confusion, and the poor financial status of local governments resulted in heavy dependence on the central government. Korea's long history of strong central government – more than 500 years of the Chosun Dynasty (1392–1910) – remains a strong influence for many

Koreans (KPCGID, 2003). Ten years have passed since autonomy. But the ratio of national to local taxes is still 80:20, nearly the same as before. Sixty per cent of the local governments have been unable to pay their staff's salaries with their local income, and among the 234 basic local governments, only 22 show incomes above 50 per cent of the financial power index (computed by dividing the governmental unit's income by its budgetary needs or obligations).

Reports of conflicts among the central, provincial and local governments, and between governments and people, have increased. Nationally mandated construction of infrastructure, such as roads, railways and nuclear power facilities, creates serious budgetary problems and increased conflicts and antipathy among stakeholders (Oh, 2004).

Because of the industrialization and modernization brought about by several economic plans under the central government system, in the early 2000s the urbanized population reached more than 80 per cent, the ratio of old people surpassed that of any other nation, and the birth rate dropped to the lowest in the world (KNSO, 2006). In the 1950s and 1960s, several short-term forest development plans for devastated forests could not be successfully accomplished because of inconsistency in implementation and budget shortages (Yoo, 1997a and 1997b). Beginning with the Third Economic Plan (1972–1976), long-term forestry was identified as a main interest of the president and the government, and the first 10-year plan for forestry began in 1973.

Forest Resources and Forestry

General status

Forests and mountains are important in Korea's history, culture and future national development. In 2003, the forest area of the Republic of Korea was around 6.42 million hectares, or approximately 64 per cent of the total land area. National forest accounts for 22.5 per cent of the forest area, and public forest owned by local governments is around 7.6 per cent. The remaining 69.9 per cent comprises small-scale private forests owned by 2.32 million people. Conifers cover 2.7 million hectares, broad-leaves 1.7 million hectares and mixed forest 1.9 million hectares (KFS, 2005b) (Table 7.1).

Administrative units that are more than 70 per cent forested and have below-average population density and agricultural land are classified as mountain village areas, or *Eup* and *Myun*. These are managed separately as 'mountain village promotion areas'. The area of mountain villages is about 4.6 million hectares. They are primarily Buddhist, with 4.7 per cent of national households and 3.9 per cent of the national population (KFRI, 2003; Yoo and Seo, 2005). At an average of 41 people per square kilometre, the population density of mountain areas is only 8.5 per cent of the national average (482 people per square kilometre).

Table 7.1 *Forest area, ownership and population*

Classification	*Land area (1000ha)*	*Forest ownership (1000ha)*				*Households (1000)*	*Population (1000)*
		Central government	*Local government*	*Private*	*Total*		
National (A)	9954	1441	491	4484	6416	14,834	48,021
Percentage		22.5	7.6	69.9	100.0		
Mountain village (B)	4569	1124	328	2302	3755	690	1870
Percentage		29.9	8.7	61.3	100.0		
Percentage (B/A)	46.0	78.0	66.9	51.3	58.5	4.7	3.9

Source: KFRI (2003)

The *Si* and *Gun* administrative units are responsible for private forests and mountain villages in their area and regard mountain villages as an important policy arena. National forests located in mountain villages are directly managed by central government foresters, but they are also strongly related with public and private forest management. The northeastern part of the country, Gangwon-do, has a high ratio of mountain villages and considerable steep mountain area.

In 2003, the domestic production of forest products amounted to about 3197 billion Won (US$1 equalled 1170 Korean Won in 2003): 103 billion Won (3.2 per cent of the total) in timber production, the rest being nuts and fruits (365 billion Won), wild vegetables (168 billion Won), medicinal plants (16 billion Won), agricultural materials (237 billion Won), mushrooms (277 billion Won), and other non-timber forest products (KFS, 2005b).

The value of the total timber stock (72.6 cubic metres per hectare in 2003) is estimated as 24.5 trillion Won, and the stock volume is the basic resource of the forest. In 2003, other values of the forest, which include such services and benefits as erosion control, outdoor recreation and wildlife protection, reached 8.2 per cent of the nation's gross domestic product (721.3 trillion Won), or about 2.6 times the combined value of agriculture, forestry and fishery products (22.8 trillion Won) and 18.4 times the value of forestry products (3.2 trillion Won). The value of the forest welfare function has increased 3.3 times since 1989. The total value, both tangible and intangible, of forest products in forest mountain village areas, however, is jeopardized by damage from insects and diseases, fire and soil erosion (KFS, 2006).

Changes in forest management by period

After the colonial period, the Korean War and social instability in the 1950s, forest volume dropped to around 6 cubic metres per hectare, and the percentage of devastated forest reached 80 per cent in the early 1960s. In 2005, the forest stock per hectare increased to around 79 cubic metres, and the forest quickly recovered its stability. However, the average volume per hectare is still lower than Asia's or the world's average. A big challenge for Korea is the management of planted forests, which occupy more than a million hectares.

The republic's First Forest Plan (1973–1978) restored 770,000ha of denuded forest and reforested more than 2,160,000ha; the Second Forest Plan (1979–1987) planted 80 commercial forest zones and reforested 325,000ha. The third (1988–1997) harmonized economic benefits and public interests and introduced the idea of sustainable forest management. In the fourth (1998–2007), the government-led reforestation programme ended, and diverse social and international needs gained attention.

Forest stabilization was thus accomplished. The planting of large plantations was completed in the mid-1980s, and illegal logging on a mass scale was stopped by the arrest of illegal loggers and the dismissal of the government employees involved. Forest protection, especially forest fire management, was also much improved by giving greater responsibility for prevention and control of forest fires to local forest officers.

The Third Forest Plan, completed in 1997, came two years after the introduction of decentralization. At that time, the average stock volume per hectare reached 50 cubic metres, and many changes occurred in forest resources management and forest land use. Planted forests were managed sustainably, forest roads constructed and forest workers organized to deal with labour shortages for forest management in rural areas. However, after local autonomy was achieved, the majority of local governments reduced the number of forest officers, merged forestry with other governmental organizations, and cut back forest-related projects. The monetary crisis in 1997 provided justification for increasing forest budgets to hire unemployed workers for large-scale forest management activities. Nevertheless, the downsizing of well-qualified human resources and the forestry organization caused difficulties in forest management under decentralization.

The government allocated 740 billion Won to forestry in its 2001 budget, and the share for forest protection, including helicopter operations and fire recovery, reached around 23 per cent of the total budget. After the mid-1990s, the forest required a larger amount of labour and budget, especially for silvicultural activities like thinning and debranching, and poor and small-scale private forest owners asked for more financial support. Local governments, however, also suffered from low financial resources and problems related to increasing population density in some areas.

Finally, the local governments requested more support from the central government, largely to deal with the frequency of forest fire outbreaks after decentralization. Forest fire management has since been one of the hottest issues between the central and local governments. In the past 10 years, particularly large forest fires have damaged resources in Gangwon and Chungnam (Table 7.2) and caused social and economic damage to the trees, soil, and by-products such as pine nuts,

Table 7.2 *Recent large forest fires*

Date	*Location*	*Burned area (ha)*	*Lost value (billion Won)*
April 1996	Gosung, Gangwon-do	3762	26
April 2000	East Coast, Gangwon-do	23,794	36
April 2002	Cheongyang, Chungnam-do	3095	6
April 2005	Yangyang, Gangwon-do	973	n/a

Source: www.foa.go.kr

mushrooms and chestnuts, as well as to those residing near the forest. The 2005 forest fire in Yangyang, Gangwon-do, caused cultural damage as well, burning a 1000-year-old temple, Naksan-sa, in the mountains.

The damage clarified the responsibility of local governments for fire prevention and suppression. Major weaknesses of local government include the huge reductions in forest officers and organizations, the failure of early countermeasures, a lack of forest roads, and inadequate thinning, with the result that forest fires spread quickly in the mountains.

Forest Fire Management Policies

Trends in forest fires

Characteristics

The Republic of Korea is highly susceptible to forest fires because of forest structure and topographic and climatic conditions. The forest is currently characterized by thick growth because of insufficient thinning and debranching, and its thick layer of fallen leaves catches fire and burns easily. Forest fires also spread rapidly in these mountainous areas (fires spread eight times faster on steep slopes than on level ground). March, April and May are the most dangerous months, characterized as they are by high winds: 68 per cent of forest fires occur in the spring, as against 22 per cent in December, January and February (Table 7.2). The causes of forest fire were mainly anthropogenic: visitors to the mountains and graves caused 49 per cent, for example, burning rice fields and farms 18 per cent, and cigarette smoking 10 per cent. Forest roads and water sources are insufficient in the mountains, and using helicopters in strong seasonal winds and at night is problematic (MAF, 2003; Lee, 2005a and 2005b).

Frequency

Since 1980, an average of 337 forest fires have occurred each year, with damaged areas averaging 2100ha per year. Forest fires increased after the mid-1990s, with the

Table 7.3 *Forest fire occurrence and damaged area*

Year	*No of fires*	*Area burned (ha)*	*Year*	*No of fires*	*Area burned (ha)*
1980	403	1218	1994	433	781
1981	252	814	1995	630	1013
1982	136	509	1996	527	5368
1983	135	919	1997	524	2330
1984	101	922	1998	265	1014
1985	165	363	1999	315	473
1986	275	3414	2000	729	25,953
1987	87	191	2001	785	963
1988	270	878	2002	599	4467
1989	294	1652	2003	271	133
1990	71	175	2004	544	1588
1991	139	429	Total	8408	52,485
1992	180	640	annual average	337	2100
1993	278	1752			

Source: KFS (2005b)

average number in 2000–2004 reaching 586 annually. The damaged area increased threefold, from 2100ha to 6620ha per year. During the same period, the number of forest fires damaging areas of more than 30ha reached 37, and the total affected area was 30,239ha. Total damage amounted to approximately 74 billion Won, including 5.5 million damaged trees (KFS, 2005a; Lee, 2005c) (Table 7.3).

Development of fire policy

1950s

There was no special forest fire policy before the establishment of the Republic of Korea in 1948. During the Korean War (1950–1953), a forest protection act in 1951 gave the responsibility of forest fire control to villages near forests and mountain areas. The provincial governor was to report forest fires in forest protection areas (1008 localities, 485,000ha) to the Minister of Agriculture. In 1958, the authority to grant permits for use of fire in forests moved from the police to the forestry office, and the forestry office took on the responsibility of forest fire management.

1960s

In the early 1960s, before the Forest Development Plan, harvesting trees damaged by forest fire was not permitted, despite an increase in applications to do so. After the 1967 establishment of the Forest Administration, forest fire prevention regulations were reissued as part of its official regulations in 1968.

1970s and 1980s

In 1970, the number of forest protection staff members increased to 1002, and forest fire prevention activities intensified. In 1971, a forest fire control system using helicopters was introduced in Korea. In 1973, the Forest Administration was moved to the Ministry of Home Affairs from the Ministry of Agriculture and Forestry for more effective reforestation and forest protection. Local forest organizations were established in all local governments – a directorate at the provincial level and a division at the district level. In the same year, implementation of the first 10-year forest development plan, designed to replant trees in denuded areas, began. Forest fire protection activities were emphasized, and forest fire alarm stages were also classified in this period. Entrance roads to forests were closed and entry into mountain areas was strictly controlled, depending on the level of forest fire threat (Table 7.4). Punishments were established for people who started forest fires, and the chief of the local government had the responsibility for forest fire management – and was held responsible when forest fires occurred. The strong measures created a perverse incentive, however, and local government officials were unwilling to report forest fires.

Table 7.4 *Classification of forest fire alarm stages*

Threat level classification	*Criteria*	*Action plan*
Caution warning (orange)	Danger index: 61–80. Humidity: 40–60%. Wind speed: below 6m/sec.	Station more than 1/6 of the forest officers on 24-hour duty. Restrict access to 30% of forest area and close 50% of hiking trails (9896km).
Danger warning (red)	Danger index: 81–100. Humidity: below 40%. Wind speed: above 7m/sec.	Station more than 1/4 of the forest officers on 24-hour duty. Restrict access to 50% of forest area and close 80% of hiking trails.

Note: The danger index ranges from 0 to 100 depending on the possibility of forest fire.
Source: www.foa.go.kr

Since the mid-1990s

When local autonomy was instituted in 1995, a new forest fire control system began, using 'public servicemen', who serve the public instead of the military, but large-scale forest fires still occurred. The Forest Service provided the legal foundation in 2001 to establish wildfire control command guidelines. In December 2004, the Forest Service made forest fires subject to guidelines for unifying command and authorization, provided a standard manual for action against forest fire risk, and established fire suppression guidelines. In November 2005, the Forest Service established integrated incident command guidelines and provided a manual under the National Security Council as a level of national disaster management between the central and local governments (KFS, 2004). The changes entailed establishment of a forest fire control and training division in the Forest Service, expansion of the research division in the Forest Research Institute, organization of a forest fire suppression team, purchase of large helicopters, many experiments in forest fire control, regulations on forest fire prevention and management for large-scale forest fires, the installation of forest fire monitoring cameras, the expansion of the forest fire control branch, and the short-term renting of gigantic helicopters.

Nevertheless, it is expected that forest fires will increase in the future because of the diminishing and aging population in rural areas being unable to monitor effectively. Other contributing factors include increases in the number of dry days and monocultures of pine in some areas. Moreover, the high number of recreational visitors makes speedy control at early stages in rural areas more difficult, and burning rice fields near the forest – a practice that is ineffective in reducing insects and diseases and not recommended by the government – increases the chances of forest fires in spring.

Current Issues in Forest Fire Management

Autonomy has sparked several contentious issues in forestry between central and local governments, including the rights to use forest land, the management of national forest and forest fire control. These conflicts are usually related to the management of valuable natural resources – land, human resources and expensive equipment. Forest fires are a particularly good example for examining the balance and concentration of power and responsibility in natural resources. Fire management requires large budgets and many human resources for prevention and suppression in private forests. In large, remote and isolated areas, it is not easy to monitor and control forest fires. Control requires advanced technology and expensive equipment, even in sparsely populated areas and areas with impoverished local governments. Coordination and cooperation between the various levels of government and the people are required to reduce the risk and damage from forest fire.

Forest fire management under decentralization

Currently, the central government manages fire projects and budgets, and the local government implements forest fire prevention and suppression in the field. However, local governments have become more interested in forest land use, even as they reduced the organizations and staff for forest management due to budget concerns (Park et al, 2004). Their interest in forest fire control has also diminished. Until 1995, the guidelines of the Ministry of Home Affairs, which controls and manages local governments, included regulations to punish provincial governors and local forest officials who failed to control fires, but such regulations are no longer effective. Since local autonomy and decentralization, the central government has lost its strong control function. The local forest officials, for their part, rely heavily on air suppression by aircraft and crews sent by the central government because of difficulty of access to the mountains and shortage of local staff. As a result, forest fires often spread out of control.

Poor financial status of local government

More effective forest fire control in the field requires that local governments be financially sound. However, the financial power index of local governments in mountain areas is lower than the national average (0.34). Among the local governments with a financial power index of less than 0.2, 79 are in mountain villages, representing 96.3 per cent of all local governments at that economic level; local governments with a financial power index under 0.2 formed 66.4 per cent of the nation's administrative units in 2002 (Table 7.5).

Table 7.5 *Financial power index of local governments*

Classification	*Nationwide*			*Local governments with mountain villages*			*Average B/A (%)*
	Si (no)	Gun (no)	*Total A (no (%))*	Si (no)	Gun (no)	*Total B (no (%))*	
FPI <0.2 (no)	12	70	82 (50.3)	12	67	79 (66.4)	96.3
FPI 0.2–0.4 (no)	21	17	38 (23.3)	11	10	21 (17.6)	55.3
FPI >0.4 (no)	41	2	43 (26.4)	17	2	19 (16.0)	44.2
Total no	74	89	163 (100)	23	77	119 (100)	73.0
Average	0.54	0.17	0.34	0.41	0.16	0.18	52.9

Source: Yoo and Seo (2005)
Note: FPI = financial power index

The entire area under some local governments is classified as mountain villages. These administrative units are particularly poor: HwaChun has a financial power index of 0.16, InJae 0.13; PyoungChang 0.25, SamChuck 0.21, YeongYang 0.06 and CheongSong 0.10. All are located in Gangwon Province, which has a particular problem with fire control and has experienced frequent large fires. Their financial condition makes it difficult to allocate sufficient budget for forest fire management, and they thus require more support from the central government.

Because of the local government's insufficient budget, logging debris related to silvicultural treatments such as thinning and debranching is also one of the serious causes of forest fires; such dry slash facilitates the expansion of fires in the mountains.

Diminished forest administration structure

Since 1998, many local governments have started to reorganize, and forest organizations were downsized in the interests of cost-effectiveness. Gangwon Province reduced the number of forest officers by over 25 per cent, from 400 to 292 people. Forest divisions merged; currently only 7 local governments have forest divisions and 11 have no forest division in their organizational chart. Such reductions diminish the capacity of local governments to perform basic functions like fire control and forest management. Often, for economic reasons, forest organizations are left as just one sub-unit with two or three officials. This kind of reform has also reduced the pride of forest officers and their sense of mission for forest management. Unifying or merging administrative units at the local level into larger-scale local governmental units would improve efficiency in natural resource management.

Insufficient infrastructure

Forest roads are important for forest management and forest fire control. The density of forest roads in Gangwon Province, at 2.37 metres per square kilometre, is just 5 per cent of the forest road density in developed countries. New forest roads are not constructed because of fears of collapse during typhoons and heavy rains. The insufficient road construction budget is another reason local governments are not building new roads.

Changes in natural and social conditions

The number of dry days has continuously increased in recent years (45 days in 1998, 78 in 1999, 96 in 2000, 116 in 2001 and 136 in 2002), perhaps because of climate change, and seasonal winds have come irregularly. The high forest volumes and flammable materials such as dry leaves increase the incidence of forest fires and expansion in scale in steep forests in the dry season. Social changes affect the forest, too: the five-day working week, instituted in 2005, has increased the number of hikers in the forest and thus the risk of forest fires at weekends and during long holidays.

FUTURE FOREST FIRE MANAGEMENT

Though they have become more frequent, large-scale forest fires are still unusual in the Republic of Korea, and many people and the media are interested and concerned. Post-fire forest recovery, however, is not easy to achieve in a short period of time. Therefore it is important for governments to prevent and suppress forest fires carefully and quickly to reduce damage to the environment and social and economic activity – as much for the people as for the forest itself. The central and local governments' guidelines for forest fire control – summarized below – clarify their respective roles and responsibilities.

Goals

The goals of forest fire control are the conservation of forest resources and forest ecosystems and prevention of damage to people's property and lives through forest fire prevention, preparation and suppression by the central and local governments. These include:

- the construction of a systematic and scientific, integrated system for forest fire control with forecasting of forest fire risk, wireless communication networks, global positioning systems and a statistical system;
- better public outreach on preventing forest fires and reducing the risk of forest fires started by people; and
- early detection of forest fires using remote monitoring systems to permit rapid response by helicopters and crews (KFS, 2005c).

Preparation

During the spring and autumn fire seasons (February 1 to May 5 and November 1 to December 15 respectively), the central government announces the forest fire danger index on TV and the homepage of the Forest Service. A forest fire control headquarters in the Forest Service coordinates with local governments' control centres and with relevant agencies such as the police, army, National Information Agency and National Park Office.

Prevention

The first step is to minimize forest fires. Precautionary activities by central and local governments include strengthening the prohibition on bringing flammable materials into the forest, controls on entering the forest, closing forest roads during the peak fire season and using a forest fire warning system determined by weather conditions. Prevention in the national forests is handled by the national forest management regional offices, and private forests are handled by local government officers. The second step involves placing restrictions on bringing flammables into forest fire outbreak areas. Third, both levels of governments provide public information using TV, newspapers, radio, magazines and leaflets. Finally, the participation of NGOs and civil society in forest fire prevention is encouraged for urban and rural peoples. Table 7.6 gives details of fire-related penalties.

Table 7.6 *Forest Law provisions regarding forest fire*

Violation	*Punishment*
Accidental forest fire	Maximum 3 years in prison
Arson of another party's forest, or of a protection forest, seedling forest, natural protection forest, experimental forest, genetically superior 'plus-tree' for silviculture or protected tree	Minimum 7 years in prison
Arson of one's own forest that burns adjacent forest	Minimum 10 years in prison
Damage (of forests and trees listed above)	Maximum 10 years in prison and maximum 15 million Won fine
Arson of one's own forest	1–10 years in prison
Other violation	Fine

Source: www.foa.go.kr

Suppression

The first step in suppression involves early detection of forest fires and quick reporting by local governments, achieved with a ground lookout system, surveillance by helicopter, a community reporting system by residents and an automatic camera surveillance system. The second step is expanding forest fire control with a specialized air task force, ground fire control organization, forest fire assistant teams in specific locations, better training and education for smoke jumpers and forest fire-fighters, and demonstration exercises for upgrading forest fire control techniques. The third step is acquiring additional forest fire control equipment – helicopters, local helicopter sheds and ground control equipment – and employing this efficiently. The fourth step is construction of an integrated scientific system using the integrated incident command system, instituted in 2005. A field command centre coordinates human resources, evacuation control and access to helicopters, depending on the size of the forest fire. Suppression by air is usually supported by the central government, but the main responsibility for forest fire suppression belongs to local governments.

Post-fire management

The responsibility for post-fire management of private forests rests with local officers. The first step is observation to identify outbreaks of fire after control measures have been taken, with the formation of a forest fire watch team after the forest fire is under control. The second step is reporting the damage to the head office, including an initial report to the Forest Service central office, frequent reports of forest fire control and a concluding report after the forest fire is under control. The third step is determination of the cause of the forest fire and follow-up measures and reports relating to illegal acts and lawbreakers.

Conclusion

During the past 50 years, social, economic and political developments have promoted autonomy and decentralization in the Republic of Korea. In the 1950s, the Republic of Korea was one of the poorest countries in the world. Social stabilization and people's environmental concerns allowed the country to complete its reforestation projects, and autonomy was reintroduced in 1995 in response to strong demands for democracy.

Since autonomy, there have been many changes, trials and errors in the nation and in forestry. Forest fires causing serious damage have been one of the serious national concerns during the past 10 years. Forest officers need to devote more time to forest fire control relative to other forest management activities. Despite their diligence, the unexpected spread of forest fires in some areas has been blamed on forest officers, rather than on forest owners or careless people.

Because the economic value of timber has increased only slowly, the growing environmental value of forests enlarges the central government's responsibility in forest fire management. An efficient, decentralized forest fire management system is an urgent need for the future. This will require systematic cooperation, clarification of the division of responsibility and power between the central and local governments, and a fundamental change in forest owners' and governments' consciousness in this new age of autonomy. The central government particularly must upgrade its fire preparation, prevention and suppression capabilities.

Local governments, too, should improve their command capability for forest fire control in the field and their citizen mobilization systems. Low budgets compound local governments' challenges in managing natural resources efficiently. The voluntary participation of stakeholders in forest fire management needs to be improved. When the economic value of private forests does not meet their expectations, forest owners have been unwilling to properly manage and protect their forests from fires.

Especially in the Republic of Korea and other decentralized countries with sparse population in mountainous areas, the modernization of forest fire control, combined with strong interest on the part of the people, is necessary to limit forest fire damage. One way to increase the efficiency of forest fire management in the Republic of Korea would be to strengthen the provincial-level fire management functions. This would allow the aggregation of small forest lands (the average forest area of local governments is around 28,000ha) into areas that could be more efficiently managed, and the sharing of fire management officials among several local governments.

REFERENCES

KFRI (Korea Forest Research Institute) (2003) 'National mountain villages survey report', Taejeon, Republic of Korea (in Korean)

KFS (Korea Forest Service) (2004) 'The report for forest fire prevention for 2004', Taejeon, Republic of Korea (in Korean)

KFS (2005a) '2004 forest fire statistics', Taejeon, Republic of Korea (in Korean)

KFS (2005b) *Statistical Yearbook of Forestry*, Taejeon, Republic of Korea (in Korean)

KFS (2005c) 'Forest fire control policy and integrated incident command guidelines in Korea', Taejeon, Republic of Korea

KFS (2006) 'Environmental value of forest in Republic of Korea', Taejeon, Republic of Korea (in Korean)

KNSO (Korea National Statistical Office) (2006) *Statistic Yearbook of KNSO*, Taejeon, Republic of Korea (in Korean)

KPCGID (Korean Presidential Committee on Government Innovation and Decentralization) (2003) 'Vision and direction of autonomy in new government', Taejeon, Republic of Korea (in Korean)

Lee, M. B. (2005a) 'Method of forest fire damage reduction by forest management', paper presented at symposium at GangNeung University, Chuncheon, Republic of Korea (in Korean)

Lee, M. B. (2005b) 'Monitoring and suppression of forest fire', ETRI (Electronics & Telecommunications Research Institute), Seoul (in Korean)

Lee, M. B. (2005c) 'Outbreak and spread of forest fire', paper presented at symposium at Bukyoung University, Busan, Republic of Korea, 14 December (in Korean)

MAF (Ministry of Agriculture and Forestry) (2003) 'Advance forest management techniques for full-functional forests against massive forest fire and alternative hazard reduction', Ministry of Agriculture and Forestry, Seoul (in Korean)

MOGAHA (Ministry of Government Administration and Home Affairs) (2006) *Statistic Yearbook of MOGAHA*, Seoul, Republic of Korea (in Korean)

Oh, J. I. (2004) 'Autonomy and local', *Local Administration Research*, vol 18, no 1, pp3–18 (in Korean)

Park, Y. K., Yoo, B. I., Sung, K. C. and Kwon, S. D. (2004) 'The status and proposals for forest land use classification system in Korea', *Korean Journal of Forest Measurements*, vol 7, no 1, pp106–115 (in Korean)

Yoo, B. I. (1997a) 'South Korea, reforestation for timber and conservation', *Journal of Forestry*, vol 95, no 10, pp38–39

Yoo, B. I. (1997b) 'In depth country study in the Republic of Korea – Status, trends and prospects to 2010', FAO, Rome

Yoo, B. I. and Seo, J. W. (2005) 'The status and development strategies of mountain villages under autonomy in Republic of Korea', *Korea Forest Research Institute Journal of Forest Science*, vol 68, pp101–120 (in Korean)

8

The Gap between Policy and Practice in Lao PDR

Yayoi Fujita and Kaisone Phengsopha

Forest policy in the Lao People's Democratic Republic (Lao PDR) had been intermittent throughout the course of modern history because of a long period of war and the country's mountainous topography, which inhibited development of a transportation network. Institutionalization of forest management began in 1989, after the First National Forestry Conference, which addressed the problem of forest degradation and called for sustainable management as part of the Tropical Forestry Action Plan.

Since the Forestry Conference, legislation on forest management has developed rapidly. One of the landmark policies, land and forest allocation (LFA), was instituted during the early 1990s.[1] It delineated village boundaries and distinguished between forest and non-forest lands. LFA also recognized the rights of villages to manage vast areas of forest, as well as the rights of individuals to use degraded forest and agricultural lands in rural areas. It was considered one of the progressive forest policies in Southeast Asia that recognized communal rights to the forest (Poffenberger, 1999).

However, after more than a decade of experience, LFA's shortfalls and constraints for sustainable forest management have become apparent. A national participatory poverty assessment points out that the policy aggravated poverty because it limited upland households' access to swidden lands (State Planning Committee, 2000). Limited access to swidden plots without alternative development has led upland farmers to shorten swidden cycles, often resulting in high labour demand for weeding and low yields of rice (Roder, 1997; de Rouw et al, 2003; Ducourtieux, 2004; Lestrelin et al, 2005). Other studies find that LFA has worsened the condition of subsistence farmers who were already socially marginalized (Lao Consulting Group, 2002; Moizo, 2004a and 2004b; Ducourtieux et al, 2005).

The current chapter examines LFA from the decentralized forest management perspective. In particular, it examines the process by which the policy was implemented and considers whether it helped build sustainable forest management at the community level. Decentralization is defined as a process of transferring and devolving power to local actors. Decentralized forest management thus seeks a transition from centrally planned resource management to that initiated by local authorities or groups of people. It is a matter of not only transferring responsibility for resource management but also allocating resources and involving local actors in the decision-making process (see also Larson and Ribot, 2004).

To analyse the process of decentralization, we need to consider what powers are being transferred and to whom. According to Ribot (2002), the ability of people to select representatives to manage resources and the existence of institutions that ensure equity, justice and efficiency are critical factors in the decentralization of resource management. Based on case studies in South Asia and West Africa, Agrawal and Ribot (2003) conclude that effective decentralization requires that empowered local actors be downwardly accountable. This requires a robust institutional mechanism that guarantees the rights of local people. At the same time, it is also important to keep in mind who is included and excluded from the process, and why. This requires inquiry into the interactions between actors and their prior power relations that subsequently affect the outcome of decentralization (Meyen and Doornbos, 2004).

The current chapter first reviews the history of LFA and the major actors involved. Second, it examines two types of allocation processes. One is undertaken by district authorities with funds from the provincial government; the other is supported by an international donor. The comparison highlights underlying problems of LFA in practice, particularly its constraints in building forest management institutions that attain the dual goals of incorporating local people's needs and protecting the environment. District authorities such as the District Agriculture and Forestry Office (DAFO), placed at the centre of LFA, are often ill equipped to pursue challenging goals because of their limited financial and human resources. The weakness of the institutional foundation means that violations cannot be prosecuted by law but instead are resolved through social and political relationships among the stakeholders.

Through review of the allocation process, this chapter shows how the policy has placed local authorities, such as DAFO and village organizations, at the centre of forest management. In Lao PDR, state control of the forest has never been strong, a situation related to the fundamental shortage of financial and human resources at the central government level to support forest management at the local level.[2] De facto decentralization was thus long-standing. However, it is also interesting to note that in upland areas, LFA was used to control shifting cultivation (often blamed for forest degradation by the country's political leaders). The weak institutional capacity of DAFO to adapt resource management plans to local conditions and monitor their impacts, as well as the still-new notion of resource tenure among the members of local communities, inhibits LFA as a means of community-based resource management.

LAND AND FOREST ALLOCATION POLICY

The Constitution of Lao PDR stipulates that the central government owns all the nation's natural resources. Nevertheless, a series of legislation during the early 1990s led to the development of the Land and Forest Allocation Programme, which recognized the rights of communities and individuals to use and manage resources. The notion of allocating forests and land to villages for management and intensive use was supported by Prime Minister Order No 3 in 1996.[3] This order called for delineating territorial boundaries at the village level and distinguishing areas of forest and non-forest. At the time of its implementation, LFA was considered one of the most progressive forest management policies in Southeast Asia, since the government not only allocated degraded forest and agricultural lands to rural farmers but also recognized the rights of communities to manage vast areas of forest (Poffenberger, 1999).

When the LFA programme was written, it also clarified responsibilities among different ministries and their line agencies. The Ministry of Agriculture and Forestry became the executive organization that supervised implementation. Its line agencies include the Provincial Agriculture and Forestry Office (PAFO) and the District Agriculture and Forestry Office.[4] Under the programme, DAFO became the primary agency that implemented the prescribed activities together with village organizations. Village organizations include administrative and political leaders of the village,[5] as well as other functional leaders, such as village forest volunteers (*pā mai xao bān* or *pā mai bān*).

In the mid-1990s, the government allocated an annual budget to DAFO through the Department of Forestry and PAFO. International donors also contributed significantly to the development of LFA. According to Daoroung (2000), by 1998 there were 26 agencies housed at the Centre of Protected Areas and Watershed Management under the Department of Forestry.[6] Bilateral agencies, including the Swedish International Development Agency (Sida) and Deutsche Gesellschaft für Technische Zusammenarbeit (GTZ), helped write the procedures and techniques. These two organizations have continued to play a crucial role in monitoring and evaluating the village-based forest management process since 2000 (Jones, 2000; Rock, 2003). Other donor projects that supported reforestation, conservation and production forest management at the village and community levels during the 1990s also required LFA as a precondition for initiating their activities (Braeutigam, 2003).

By 2003, LFA was being conducted in more than 5400 villages across the country, accounting for almost 50 per cent of the total number of villages in Lao PDR (Soulivanh et al, 2004). An article in *Vientiane Times* on 19–22 January 2001 claimed greater success, with 7000 villages. The article also suggested that LFA had helped control the expansion of shifting cultivation and alleviate rural poverty in upland areas of Lao PDR. In the Forest Strategy 2020, issued in 2005, the Ministry of Agriculture and Forestry also claimed success for LFA in reducing the area under shifting cultivation, indicating that the political priority was controlling the spread of shifting cultivation as a way to protect forest.

However, the actual results of the allocation on the ground are mixed. Although some villages were willing to adopt the policy, others continued to convert forest lands into agricultural lands despite the zoning introduced by LFA (Nakatsuji, 2004). Some upland villages were displaced because of LFA, and their residents, lacking alternatives, suffered severe poverty (State Planning Committee, 2000). LFA was generally useful for villages that were located near commercially valuable resources and could organize themselves to claim their exclusive legal rights (Fujita et al, 2005). It was not so meaningful for those villages located far from the resources they had traditionally used. The following section examines the outcomes, based on different studies.

Policy in Practice

After more than a decade of experience, several studies suggest gaps between LFA policy and its achievements. Table 8.1 summarizes the implications of LFA for the environment, livelihoods, displacement and resource tenure.

One of the most frequently noted effects of LFA on the local environment is its restrictions on swidden fields. According to a study conducted by GTZ, more than 80 per cent of all village forest was categorized into conservation and protection forests after LFA. Through the allocation process, swidden-fallow lands were incorporated into forest lands (Soulivanh et al, 2004). Although the swidden fields decreased, studies by Thongmanivong et al (2005) and Thongmanivong and Fujita (2006) indicate an overall trend of increasing secondary forest, particularly during the late 1990s and 2000. However, increased forest cover does not mean that forest degradation has stopped. Most of the forests that are recovering are fragmented and have little commercial value.

Studies by Roder (1997) and Lestrelin et al (2005) point out direct effects on fallow cycles as a result of regulating the expansion of areas in which shifting cultivation can be practised. In some places, fallow cycles have been reduced from 10 or 12 years to 3 or 4 years. The shortening of the fallow period has caused serious infestations of weeds in most places and increased household labour requirements during the crop cycle (Roder, 1997; UNDP, 2001; Lestrelin et al, 2005). It has also led to soil erosion and loss of fertility, further depressing yields of upland rice and compromising household food security (de Rouw et al, 2003; Ducourtieux, 2004; Lestrelin et al, 2005).

Table 8.1 *Implications of LFA*

	LFA impacts	*Studies*
Shifting cultivation	Reduced access to swidden lands resulting in shorter fallow cycle Land degradation Reduced rice yields	Roder (1997), UNDP (2001), de Rouw et al (2003), Ducourtieux (2004), Lestrelin et al (2005)
Forest cover	Increase of secondary forests	Thongmanivong et al (2005), Thongmanivong and Fujita (2006)
Forest resources	Restricted access to non-timber forest products Reduction of commercially valuable non-timber forest products	Takeda (2003), Yokoyama (2004)
Livelihoods	Increased dependence on non-timber forest products, cash crop production and wage labour	Raintree and Soydara (nd)
Relocation	Increased internal migration Increased vulnerability	SPC (2000), Evrard (2003), Vandergeest (2003), Moizo (2004a and b), Baird and Shoemaker (2005)
Resource tenure	Conflicting interests of state and communities over resources Increased competition over commercially valuable resources	Eggertz (1996), Evrard (2003), Tubtim and Hirsch (2005), Fujita and Phanvilay (2008)

Still other studies indicate that the increased emphasis on forest conservation and the loss of swidden-based forest management systems have minimized local people's access to non-timber forest products. Studies by Takeda (2003) and Yokoyama (2004) point out the versatility of the traditional swidden-fallow system, which allows the sustainable harvest of such non-timber forest products as cardamom and benzoin along with upland rice cultivation. As swidden and fallow areas are restricted, old fallow areas where non-timber products are harvested become subject to conservation, and plants that were part of the long-term farming system in the upland areas are crowded out by forest regrowth.

Furthermore, lack of alternative livelihoods after LFA and the loss of food security have forced some upland villages to relocate (State Planning Committee, 2000; Vandergeest, 2003; Evrard, 2004; Moizo, 2004a and 2004b; Lyttleton et al, 2004; Baird and Shoemaker, 2005). As a result of such relocations, the population is increasingly concentrated in lowland areas along the roads, thereby increasing competition for resources in certain areas across the country. The pressure on resources is also on the rise as rural communities become integrated into the market economy and the commercial value of resources increases.

In the meantime, the still-new concept of exclusive ‘rights’ and the lack of judicial institutions that guarantee local people’s legal rights also mean that, despite LFA, the status of the communities’ and individuals’ rights to resources remains uncertain. Tubtim and Hirsch (2005) contend that local communities have continued to access resources for subsistence purposes, following their customary practices, even after LFA is implemented.

The effects of LFA, summarized in Table 8.1, also highlight gaps between theory and practice. The first gap occurs because legal categories for forest and land stipulated by the Forest Law[7] differ from customary forest distinctions. LFA applies five forest categories, zoning each forest as a contiguous unit within the village boundary. Management rules for each forest area also follow the principles underlined in the Forest Law.

Although LFA generally invites local people’s participation in the forest delineation process, the quality of participation is highly questionable, since the objective of LFA is perceived to be to zone forest areas. The result is a simple sketch map (Figure 8.1) and a management plan which does not differ from one village to the other. The process disregards the multiple values of forests and local communities’ complex resource tenure systems. In sum, LFA makes local forest management practices ‘legible’ in the eyes of the local authority, but the way local people perceive forests may differ.

The second gap is created when an exclusive village boundary is delineated, isolating one village from the next. Such boundaries are administrative, making each

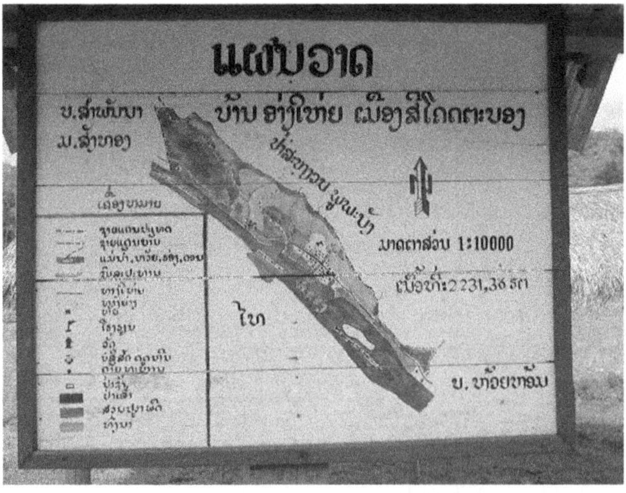

Figure 8.1 *Typical LFA map*

The customary village territory included a vast forest area, which was incorporated into the national conservation forest in 1993. The village is now left with secondary forest, paddy field, agricultural land and a residential area.

village responsible for the territorial space and management of the resources within it. At the same time, they enforce exclusivity of resource use and management to a village; in traditional practices, in contrast, neighbouring villages have shared the use of resources. Although territorial boundaries make it easier for the local authorities to oversee resource management practices, they have also sparked new conflicts between communities. The conflicts are intensified when claims over scarce and commercially valuable resources are in question.

A third gap concerns who participates in the planning process and how they participate. All LFA actions involve local authorities, such as DAFO, and village organizations. The village organization is a representative body of the village, but it does not necessarily represent all groups of resource users in the village, such as women and men who enter the forest daily to collect food. Although members of organizations such as the Women's Union are part of the village organization and attend village meetings, their role in decision-making remains minimal. Conscious effort is required to identify different groups of resource users and their interests. Time to facilitate discussions with groups of stakeholders during the resource management planning process is also needed.

The way in which members of the village organization participate in LFA is often affected by existing social relationships. Those with more power – determined by years of residence, proximity to resources, links with political leaders, and ability to mobilize capital and labour – within the village are often able to use LFA to their advantage and make legal claims to lands and resources, while socially marginalized people have limited opportunities to negotiate their access to productive lands. Nor are they fully aware of the potential benefits they can gain through LFA, since they are generally absent during the planning process.

Finally, the last gap is in local authorities' capacity to adapt LFA to people's changing relationship with natural resources. Given the rapid integration into the market economy and the dynamic movement of people, local people's use of resources has been changing over the past decade. Lack of financial resources is often cited as the major constraint; however, the weakness of the institutional foundations of DAFO inhibits the continuity of LFA and its follow-up activities. Local authorities are also unable to arbitrate emerging conflicts and balance the need to protect the environment with local people's needs for resources.

In the following section, we compare two types of allocation processes to further explore the constraints of implementing the policy and factors that are crucial for decentralized forest management.

Comparison of LFA Processes

Every allocation involves district authorities and the village organization. However, the government-led LFAs differ from donor-sponsored LFAs.

Government-led LFAs

Under MAF Instruction No 0822, LFA was broken down into eight procedures (Table 8.2). DAFO is the lead agency that works with village organizations. In most cases, the LFA team consists of the DAFO staff, which carry out Steps 1 to 6 as described in Table 8.2. Generally, activities are conducted within one to two weeks with an average budget of US$100 per village (Soulivanh et al, 2004).[8] The budget is allocated through the district and provincial government offices based on an annual plan submitted by DAFO. However, the budget is often insufficient to cover planned activities, leaving DAFO to implement only the minimal procedures of LFA.

Table 8.2 *Original LFA procedures, 1996*

Procedures	*Activities*
Step 1	Basic preparation
Step 2	Village consultation
Step 3	Village data collection: forest survey, household land-use survey
Step 4	Village meeting
Step 5	Land survey
Step 6	Land-use planning
Step 7	Extension
Step 8	Follow-up: monitoring and evaluation

Source: Ministry of Agriculture and Forestry (1996)

Financial resources are not the only underlying constraint. DAFO is also faced with staff shortages. Staff have limited opportunities for formal training, and thus DAFO has generally limited capacity to pursue participatory resource management planning. At the same time, DAFO represents the state, and its policy intervention must comply with communist party resolutions, which take a critical view of shifting cultivation. This has led DAFO to focus on controlling the expansion of shifting cultivation and applying the forest categories stipulated by the Forest Law. The actual conditions of local communities and the high variability of shifting cultivation and upland agricultural practices are not fully integrated into most of the village resource management plans.

In most instances, DAFO-led LFAs end at Step 6 because of budget constraints. Villages therefore have designated forest areas and management plans. Budgets for extension activities (Step 7) are submitted separately through DAFO, often without any direct link to the management plans developed during LFA. The lack of systematic record-keeping also inhibits consistency of resource management planning at the district and village levels after LFA.

Donor-led LFAs

International donors also supported LFA throughout the 1990s, working with different levels of government authorities. Donor-sponsored LFAs generally pay more attention to checking the accountability of local authorities, such as PAFO and DAFO, to the local communities.

Although the general procedure does not significantly differ from the government-led interventions, donor-led LFAs spend more time on the procedures, encouraging active local participation in the resource boundary demarcation and planning process (Evrard, 2004). Most donor-led LFAs identify various stakeholders within the group and facilitate dialogue between stakeholders to discuss their problems. Donor-led LFAs are less concerned about financial resources and more concerned about developing a participatory process, particularly allowing the villagers to raise their concerns during the resource management planning.

Donor-led LFAs also provide training for local authorities, such as DAFO, prior to implementation of the LFA. This includes training on participatory process, government legislation and so forth. In some instances, donors provide training for village organizations in forest inventory, resource management planning, community fund management and other skills. In all training, emphasis is placed on facilitating a participatory decision-making process, involving local authorities and resource users.

Donor agencies often monitor the procedures throughout the LFA process and make necessary adjustments to ensure local people's rights to resources. They also pay more attention to strengthening local institutions at both village and district levels (Litz, 2000). Agencies such as Sida and GTZ have particularly focused on building local authorities' capacity to keep systematic records throughout LFA (Table 8.3). Donors with long-term commitments have also supported DAFO in

Table 8.3 *Revised LFA procedures*

Procedures	*Activities*
Step 1	Preparation
Step 2	Village boundary delineation and land-use zoning
Step 3	Village data collection
Step 4	Land-use planning
Step 5	Forest and agricultural land allocation decisions
Step 6	Field measurements of agricultural lands
Step 7	Forestry land agreements and transfer of rights to villagers
Step 8	Land-use planning and recording information on land allocation
Step 9	Agricultural land allocation records
Step 10	Follow-up: monitoring and evaluation

Source: Leuangkhamma et al (2001)

following up on management plans, as well as introducing agricultural extension projects after LFA.

One significant difference between government-led and donor-led allocations is that the donor agencies provide financial resources and technical services to local authorities. Another important aspect is that donor agencies often act as facilitators and monitors to ensure that local people's rights are being protected by the local authorities. However, the iterative resource management planning process ends when the donor-led projects come to a close, and thus their impact may not be long-lasting. The extent to which donor agencies become involved in facilitating the participatory process also depends on their commitment to promoting decentralized resource management and institutional capacity-building.

Lessons learned from comparison

A simple comparison between government-led and donor-led allocation processes highlights underlying constraints for LFA in Lao PDR. Because the notion of exclusive resource boundaries and rights is still very new, DAFO and village organizations often lack clear understanding of government legislation, including how customary resource-use practices can be accommodated in formal forest management institutions. As pointed out by Jones (2000), local government staff involved in LFA often prioritize the classification of forests into categories stipulated by the Forest Law without considering the diverse ways resources have been used by the community. As a result, many swidden-fallow lands have been designated as forest areas and cannot be converted into non-forest land uses.

The nature of local participation is also limited in government-led LFAs, which are often implemented so as to minimize the cost and thus tend to undermine the participatory process, instead imposing simplified resource management practices that fail to capture the needs of local communities to use resources. The village is isolated as an administrative unit without consideration of the history of local communities and their communal resource-use practices.

The comparison of donor-led and government-led LFAs also reveals the importance of supervision during and after the process. A systematic record-keeping system is needed to sustain the iterative process of developing a local resource management framework after LFA – not only for institutionalizing collective forest management, but also for securing private rights to land and resources. It is also important to recognize that LFA is not a one-time action but a continuous process that requires reviewing old land-use and management plans, adapting management practices to the emerging needs of the local communities, and ensuring the protection of the environment beyond a single village.

Finally, the comparison illuminates the importance of a supervisory body that ensures the rights of communities and individuals. This was one of the important roles played by international donors.

Balancing Needs and Adapting to Change

Although state control of forests was never strong in Lao PDR, the LFA policy transferred the management responsibilities from the Department of Forestry to DAFO and to village organizations. During the 1990s financial resources were also allocated by the Department of Forestry to PAFO, and then to DAFO. However, PAFO no longer receives its LFA budget from the Department of Forestry. Instead, the provincial government allocates budget for PAFO, which is then distributed to DAFO. This transition occurred in early 2000 as part of the central government's effort to decentralize the political structure. However, it also means that decisions to allocate funds to LFA are now dependent on the provincial governor's office. The budget for LFA in general has declined since 2000, and thus funds for new LFAs and follow-up activities are limited.

LFA has helped institutionalize forest management in villages, but it has not empowered all members of the communities to contribute in the resource management planning process. It also has not fully guaranteed resource tenure of communities or individuals against the will of powerful individuals and organizations. Anybody with the right connections with political power at the district, provincial or national level can override local resource management agreements. Under the generally weak institutional and regulatory body of local authorities, claims for productive resources continue to be influenced and defined by existing social relationships. This not only makes resource tenure uncertain but also limits the legal rights of communities to use and manage their resources, even after LFA. Monitoring of resource tenure and the state of resources in local villages after LFA – in particular, the accountability of village organizations and local authorities – is therefore needed to ensure that both the environment and people's rights to resources are being protected.

Delineating resource boundaries and writing collective management plans alone do not automatically improve community-based natural resource management. As Ireson (1995) points out, collective management of forests is much more difficult to sustain than other forms of collective resource management such as irrigation, because its vast scale means high transaction costs for community members. Successful cases of collective forest management usually involve a community with a defined interest and commitment to forest management (McKean, 1992; Meyen and Doornbos, 2004). We can draw some lessons from places in Lao PDR where villagers organized themselves into, for example, a bitter bamboo collection and sales group and a production forest management group. These success stories indicate the importance of having a common interest and being able to generate benefits for both individuals and the community. The benefits increased people's confidence in the group and its rules regulating resource-use practices.

Finally, increasing integration of rural communities into the market economy is significantly influencing how local people access and use resources. Forming production groups is one way of coping with such increased integration. However, when the expected benefits from commercial agriculture become higher than the benefits generated from the production groups, rural farmers often find themselves in competition with one another to claim valuable resources and capture the new economic opportunities. Commercialization of agricultural production particularly affects access to lands as families begin to claim exclusive rights to fields. Those who can mobilize capital and labour are among the first to make claims on productive resources. Although privatization is the ultimate form of decentralization, the current drive to engage in commercial agricultural production puts at risk communal resources that not only have environmental value but also provide food and material sources for households in times of need – particularly those families unable to capture the new economic opportunities.

LFA was intended to institutionalize village-based resource management in Lao PDR; however, it is not a panacea for sustainable community-based natural resource management. Management responsibilities were decentralized to village organizations, but local people's rights over communal and private lands remain uncertain. No regulatory body exists to monitor agreements based on LFA to ensure their effectiveness. Delineating resource boundaries and mapping is not an end in itself. The LFA programme should continue to support the dialogue between stakeholders to balance environmental protection with emerging needs for development on the one hand, and collective resource management and privatization on the other.

Notes

1 In the early 1990s, LFA was referred to as *nanyobāi kān bēng din bēng pā*, which means policy on land and forest zoning. However, this began to change into *nanyobāi kān mōp din mōp pā*, which implies more emphasis on allocation of land and forest.

2 From the late 1970s to the mid-1980s, state forest enterprises were established in Lao PDR. Although these state companies had the exclusive rights to log, forest areas generally remained open to local farmers.

3 The Prime Minister's Order on the Continuation and Expansion of Land Management and Land and Forest Allocation (*kham sang nēnam wā duai kān soup tō phankhanyāi viak gān chatsan thī din le mōp din mōp pa*) entered into force on 25 June 1996.

4 DAFO was renamed the District Agriculture and Forestry Extension Service Office (DAFEO) in 2005.

5 Community organizations include the Youth Organization and the Women's Union.

6 Seven projects were supported by bilateral agencies (GTZ, JICA, Sida and DANIDA), nine were supported by multilateral agencies (the World Bank, Asian Development Bank, UN FAO, UN Environment Programme and UN Development Programme), ten were supported by international NGOs, and four were supported by volunteer organizations (Daoroung, 2000).

7 The Forest Law was promulgated in 1996.

8 Where there is a shortage of DAFO staff, civil servants from other government offices at the district level join the LFA team.

REFERENCES

Agrawal, A. and Ribot, J. (2003) *Analyzing Decentralisation: A Framework with South Asian and West African Environmental Cases*, World Resources Institute, Washington, DC

Baird, I. G. and Shoemaker, B. (2005) *Aiding or Abetting? Internal Resettlement and International Aid Agencies in the Lao PDR*, Probe International, Toronto, Canada

Braeutigam, D. (2003) *Community Based Forest Management in Cambodia and Laos: Frame Conditions, Selected Examples and Implications*, MRC-GTZ Cooperation Programme/Cambodian–German Forestry Project, Phnom Penh

de Rouw, A., Kadsachac, K. and Gay, I. (2003) 'Four farming systems: A comparative test for erosion, weeds and labour input in Luang Prabang region', *Juth Pakai*, vol 1, pp14–22

Daoroung, P. (2000) 'Global actors, local forests: Power plays within forest management in southern Lao PDR', thesis, University of Sydney, School of Geosciences, Division of Geography, Sydney, Australia

Ducourtieux, O. (2004) 'Shifting cultivation and poverty eradication: A complex issue', paper presented at the Conference on Poverty Reduction and Shifting Cultivation Stabilization in the Uplands of Lao PDR, 27–30 January 2004, Luang Prabang, Lao PDR, National Agriculture and Forestry Research Institute (NAFRI), Vientiane, Lao PDR

Ducourtieux, O., Laffort, J. R. and Sacklokham, S. (2005) 'Land policy and farming practices in Laos', *Development and Change*, vol 36, no 3, pp499–526

Eggertz, D. (1996) 'Tenure and sustainable use of the forest land in Lao PDR', Master's thesis, University of Uppsala, Faculty of Law, Uppsala, Sweden

Evrard, O. (2003) 'Land allocation and land titling in Laos: Notes on implementation and social impacts', UN Food and Agriculture Organization (FAO), Vientiane, Lao PDR

Evrard, O. (2004) 'The implementation of land allocation and land titling in Laos: Impacts on land access and rural livelihoods', abstract, Sustainable Development Department, FAO, Rome

Fujita, Y and Phanvilay, K (2008) 'Land and forest allocation in the Lao People's Democratic Republic: Comparison of case studies from community-based natural resource management research', *Society and Natural Resources*, vol 20, no 2

Fujita, Y., Vongvisouk, T., Chantavong, H. and Chanthaleunnavong, S. (2005) 'Decentralized forest management in production forest in Central Laos Forest, Dong Phousi Production Forest and Dong Sithuoane Production Forest', in P. Durst, C. Brown, H. D. Tacio and M. Ishikawa (eds) *In Search of Excellence: Exemplary Forest Management in Asia and the Pacific,* Food and Agriculture Organization of the United Nations, Regional Office for Asia and the Pacific, and the Regional Community Forestry Training Center for Asia and the Pacific (RECOFTC), Bangkok

Ireson, W. R. (1995) 'Village irrigation in Laos: Traditional patterns of common property resource management', *Society and Natural Resources*, vol 8, pp541–558

Jones, P. (2000) 'Land use planning and land allocation experiences, lessons, and issues from the Lao Swedish Forestry Program Phase 4', paper presented at the NAFRI Research Strategy Development Workshop, 8–11 May 2000, Dongdok, Vientiane, Lao PDR

Lao Consulting Group (2002) *Existing Land Tenure and Forest Lands Study*, Ministry of Finance, Department of Lands, Vientiane, Lao PDR

Litz, V. (2000) 'Certification and local forest management: The FOMACOP experience in the Lao PDR', *Asia-Pacific Community Forestry Newsletter*, vol 13, pp41–45

Larson, A. M. and Ribot, J. C. (2004) 'Democratic decentralization through a natural resource lens: An introduction', *Development Research*, vol 16, no 1, pp1–25

Lestrelin, G., Giordano, M. and Keohavahn, G. (2005) 'When "conservation" leads to land degradation: Lessons from Ban Lak Sip, Laos', IWMI Research Report 91, International Watershed Management Institute, Colombo, Sri Lanka

Leuangkhamma, T., Sysomvang, S. and Jones, P. (2001) *Procedures and Methods for Land Use Planning and Land Allocation*, Lao–Swedish Forestry Programme, Land Use Planning Unit, Forestry Inventory and Planning Centre, Vientiane, Lao PDR

Lyttleton, C., Cohen, P., Rattanavong, H., Thongkhamhane, B. and Sisaengrat, S. (2004) *Watermelons, Bars and Trucks: Dangerous Intersections in Northwest Lao PDR: An Ethnographic Study of Social Change and Health Vulnerability along the Road through Muang Sing and Muang Long*, Macquarie University and Lao Institute for Cultural Research, Vientiane, Lao PDR

McKean, M. A. (1992) 'Management of traditional common lands (*iriaichi*) in Japan', in D. Bromley (ed) *Making the Commons Work: Theory, Practice and Policy*, International Center for Self-Governance, San Francisco, CA

Meyen, W. and Doornbos, M. (2004) 'Decentralizing natural resource management: A recipe for sustainability and equity?', *European Journal of Development Research*, vol 16, no 1, pp235–254

Ministry of Agriculture and Forestry (1996) 'Instructions regarding land and forest allocation for management and use', Ministry of Agriculture and Forestry, Vientiane, Lao PDR

Moizo, B. (2004a) 'Implementation of the land allocation policy in Lao PDR: Origins, problems, adjustments and local alternatives', paper presented at the Conference on Poverty Reduction and Shifting Cultivation Stabilization in the Uplands of Lao PDR, 27–30 January 2004, Luang Prabang, Lao PDR, National Agriculture and Forestry Research Institute (NAFRI), Vientiane, Lao PDR

Moizo, B. (2004b) 'Kmhmu responses to the land allocation policy: A case study from the Lao PDR', paper presented at Geography Conference, Khonkhaen University, Khon Khaen, Thailand, 29 November–2 December

Nakatsuji, S. (2004) 'Raosu yakihata sanson ni okeru kankinsakumotsu saibai jyuyougo no tochi riyou [Changes in slash-and-burn agriculture after the introduction of cash crops in the hilly areas of Laos: A case study of number 10 village, Xiengngeun District, Luang Prabang]', *Jinbun Chiri [Human Geography]*, vol 56, no 1, pp1–21

Poffenberger, M. (ed) (1999) *Communities and Forest Management in Southeast Asia: A Regional Profile of the Working Group on Community Involvement in Forest Management*', IUCN, Gland, Switzerland

Raintree, J. and Soydara, V. (nd) 'Human ecology and rural livelihoods in Lao PDR', National Agriculture and Forestry Research Institute, Vientiane, Lao PDR, unpublished draft 2001

Ribot, J. C. (2002) *Democratic Decentralization of Natural Resources: Institutionalizing Popular Participation*, World Resource Institute, Washington, DC

Rock, F. (2003) 'Improvement and adaptation of the land use planning/land allocation process in the RDMA program areas of Luang Namtha and Bokeo', GTZ, Vientiane, Lao PDR

Roder, W. (1997) 'Slash-and-burn rice systems in transition: Challenges for agricultural development in the hills of Northern Laos', *Mountain Research and Development*, vol 17, no 1, pp1–10

Soulivanh, B., Chantalasy, A., Suphida, P., Lintzmeyer, F. and Rock, F. (2004) *Study on Land Allocation to Individual Households in Rural Areas of Lao PDR*, German Technical Cooperation Sector Project Land Management, Vientiane, Lao PDR

State Planning Committee (SPC) (2000) *Poverty in the Lao PDR: Participatory Poverty Assessment (PPA)*, SPC, Vientiane, Lao PDR

Takeda, S. (2003) 'Land allocation programme in Lao PDR: The impacts on non-timber forest products and livelihoods in marginal mountainous areas', paper presented at the International Symposium on Alternative Approaches to Enhancing Small-Scale Livelihoods and Natural Resources Management in Marginal Areas: Experience in Monsoon Asia, 29–30 October 2003, The United Nations University, Tokyo

Thongmanivong, S. and Fujita, Y. (2006) 'Recent land use and livelihood transition in northern Laos', *Mountain Research and Development*, vol 26, no 3, pp237–244

Thongmanivong, S., Fujita, Y. and Fox, J. (2005) 'Resource use dynamics and land-cover change in Ang Nhai Village and Phou Phanang National Reserve Forest, Lao PDR', *Environmental Management*, vol 36, no 3, pp382–393

Tubtim, N. and Hirsch, P. (2005) 'Common property as enclosure: A case study of a backswamp in southern Laos', *Society and Natural Resources*, vol 18, pp1–20

UNDP (United Nations Development Programme) (2001) *National Human Development Report Lao PDR 2001: Advancing Rural Development*, United Nations Development Programme, Vientiane, Lao PDR

Vandergeest, P. (2003) 'Land to some tillers: Development-induced displacement in Laos', *International Social Science Journal*, vol 175, pp47–56

Yokoyama, S. (2004) 'Forest, ethnicity and settlement in the mountainous area of northern Laos', *Tonan Ajia Kenkyu [Southeast Asian Studies]*, vol 42, no 2, pp132–156

9

Consultative Planning for Effective Forest Governance: Case Studies from Malaysia and Indonesia

Efransjah, Petrus Gunarso, Che Hashim bin Hassan, and Kresno D. Santosa

Two projects in the Asia-Pacific region – one in Malaysia and one in Indonesia – are promoting the implementation of sustainable forest management concepts and also working towards the implementation of the UN Convention on Biological Diversity and the Ramsar Convention, to which both countries are signatories.

Malaysia has about 1.45 million hectares of peat swamp forest. Some 200,000ha remains in Peninsular Malaysia, most of it in a single, nearly contiguous area in the state of Pahang. To manage the remaining peat swamp forest resources in the state, the Forestry Department, in collaboration with UN Development Programme/ Global Environment Facility (UNDP/GEF) and Danida, has undertaken a five-year project focusing on the Southeast Pahang Peat Swamp Forest. The project, called Conservation and Sustainable Use of Tropical Peat Swamp Forests and Associated Wetland Ecosystems, began in 2002 and aims to establish integrated management plans to ensure the sustainability of the forest resources in the area. This initiative is being taken by the Pahang Forestry Department and the Forest Research Institute Malaysia, under the auspices of the Ministry of Natural Resources and Environment.

The project in Indonesia involves a tropical rainforest, the Malinau Research Forest in the District of Malinau, East Kalimantan (the Indonesian part of the Island of Borneo). The district covers 4.2 million hectares, and more than 80 per cent of it is relatively pristine forest. To promote conservation and sustainable use of the forest, the government of Indonesia, represented by the Directorate General of Forestry Research and Development (FORDA), and the Center for International

Forestry Research, in collaboration with the local government of Malinau, have begun to address the challenges of decentralization. A recently adopted decentralization law requires adjustment and adaptation for the implementation of laws and regulations, particularly for the implementation of integrated spatial and land-use planning at the district or local level. The goal is to develop integrated natural resource management planning for tropical rainforests and serve as a model for village, forest management unit and district-level management plans. Funding for this project comes from the International Tropical Timber Organization.The four-year project in Malinau concluded in 2006.

This chapter presents lessons learned and describes experiences in developing integrated management plans and spatial planning in two decentralized government settings. In Pahang, Malaysia, the issue was bringing the national commitment to biodiversity conservation and sustainable forest management down into the local governance system; in Malinau, Indonesia, the challenge was bringing local initiatives up into national guidelines and getting a national commitment to implement district conservation concepts. Participatory and multidisciplinary approaches were adopted in both initiatives, and similar challenges, to different degrees, were encountered, particularly 'attitude change' and 'business as usual' on the ground. Strong commitment at all government levels was indispensable, as were partnerships with local stakeholders.

Project Sites

Pahang

The Southeast Pahang Peat Swamp Forest is located in Pekan District, in the State of Pahang, Peninsular Malaysia, and with approximately 160,000ha is by far the largest intact peat swamp forest in mainland tropical Asia (Figure 9.1). More than half of the area, or about 87,045ha, is located within four production forest reserves. The western part is a developed, mostly agricultural area; the eastern part is mainly degraded forest land with limited potential for commercial development.

Topographically, the area is generally flat; the highest point is an isolated hill, Bukit Bangkong, 60 metres above mean sea level, in the Pekan Forest Reserve. The area consists of recent deposits (Paramanathan, 2005), which can be divided into three broad geomorphic units – coastal beach ridges and swales, the inland swamp basin and the Sungai Pahang floodplain.

In terms of soil, the inland swamp consists almost entirely of organic deposits with two common soil types, Gali Series and Gondang Series. These peat soils are defined by the presence of shallow to moderately deep (50–150cm) to deep (>150cm) partly decomposed organic soil material or hemic materials overlying riverine or colluvial clays (Paramanathan, 2005).

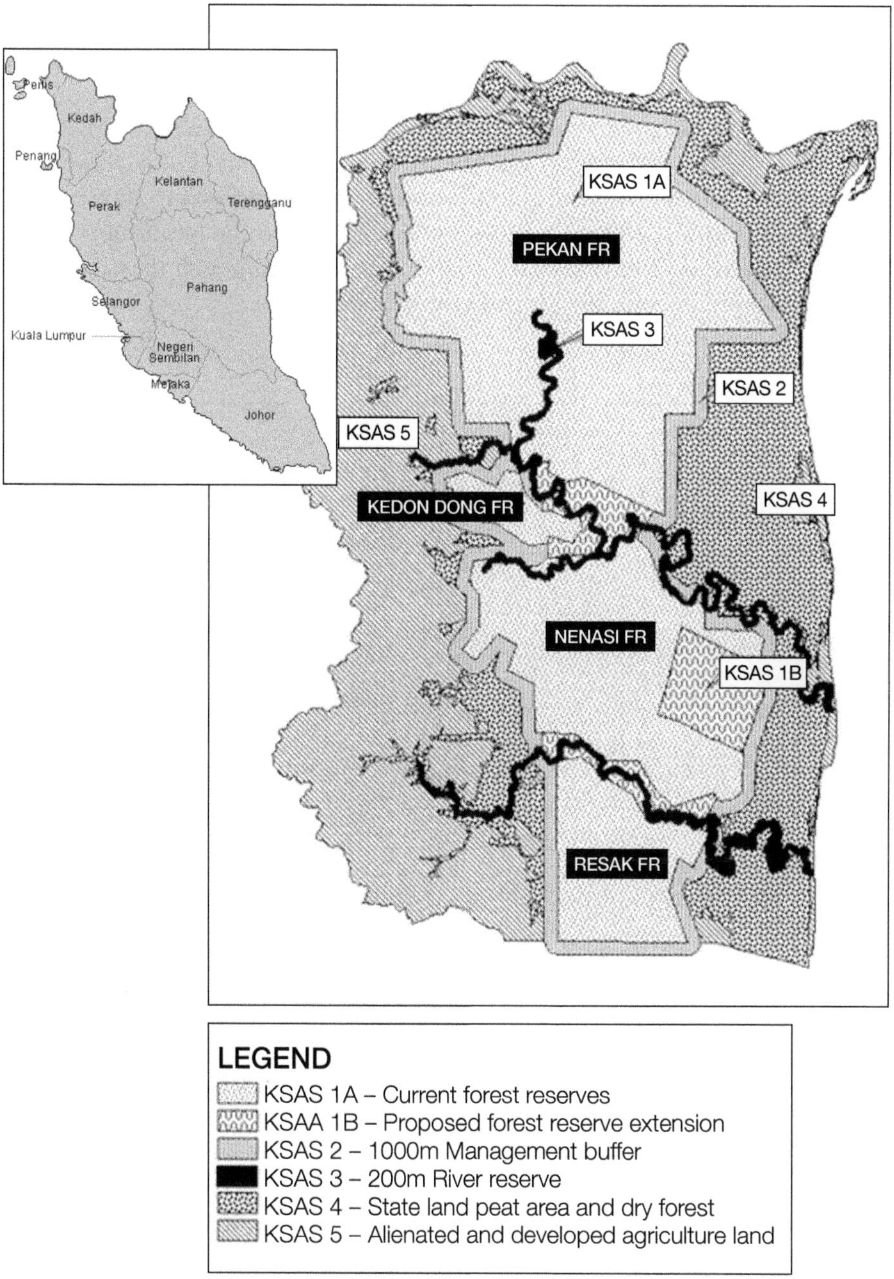

Figure 9.1 *Southeast Pahang and its management zones*

Malinau

The Malinau District in East Kalimantan was established in 1999 when a country-wide government reform introduced a highly decentralized system. During the process of decentralization, large districts were divided into two or more new districts. The political boundaries, however, were often drawn according to ethnic and historical lines, rather than watersheds. Where rivers are the only means of transportation, such divisions have created a problem in balancing conservation and development. The District of Malinau is located in the headwaters of the Malinau, Kayan and Tubu rivers. It is landlocked, connecting with the neighbouring districts only through the Sesayap river (Figure 9.2).

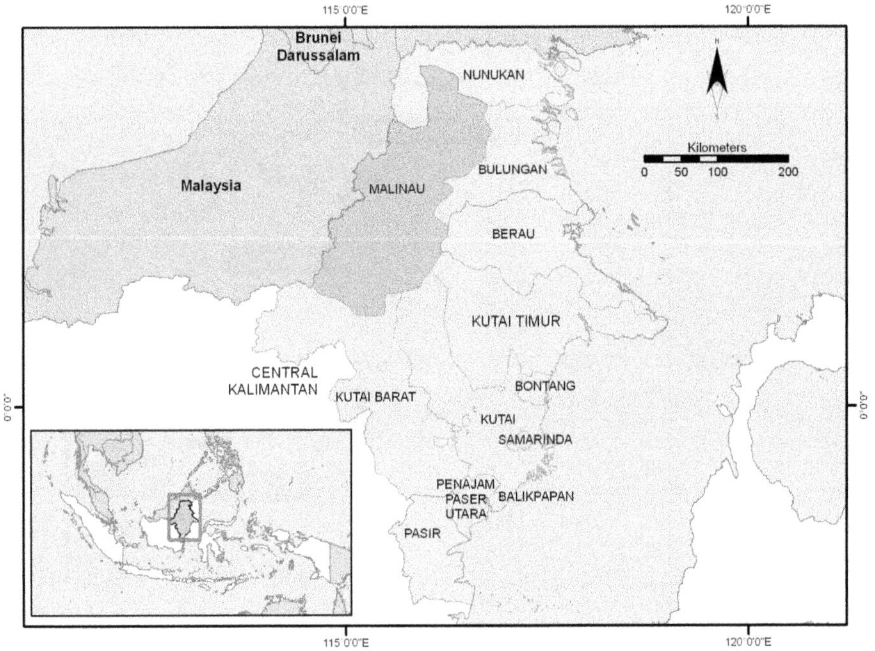

Figure 9.2 *Malinau District, East Kalimantan*

The population density of Malinau is considered very low. Although Indonesia is the world's fourth most populous country, the population density in Malinau on average is less than one person per square kilometre, largely a result of its remoteness and isolation, with vast forested areas, rugged topography and poor soil fertility. Soil surveys conducted in the upper Malinau watershed indicate that the area is not suitable for large-scale agriculture (Sheil and Basuki, 2006). In addition, the area has limited access to neighbouring districts and the highly populated and politically important island of Java.

The Need for Integrated Planning

Pahang

Under Malaysia's federal constitution, forests and land use fall within the jurisdiction of the states. Each state can enact laws, formulate forest policy and manage its forest, in line with the National Forestry Act (1984). The National Forest Council, established in 1971 and chaired by the Deputy Prime Minister, serves as a forum for the federal and state governments to discuss and resolve forest policy administration and management, including the determination of the annual timber harvest. The federal government also provides advice and technical assistance, maintains experiment stations, and funds research and training.

Approximately 80 per cent of the forest reserves in the Southeast Pahang Peat Swamp Forest is still pristine, whereas most of the state forests are in various stages of secondary succession. The growing pressure on land in the area makes an integrated management plan necessary to facilitate informed decision-making by the state. The objective of planning for the area is to ensure conservation of biological diversity and sustainable use of forest resources while maintaining ecological integrity and hydrological functions. The Southeast Pahang Peat Swamp Forest is especially important to preserve, since it represents the largest remaining intact forest of this type in mainland tropical Asia.

Peat swamp forests are specialized ecosystems with a rich variety of habitat types that support unique biological diversity. It is anticipated that by protecting the ecosystems of the area, species richness can be maintained. Thus, protection of representative habitat, flora and fauna is an important objective.

The four forest reserves in the Southeast Pahang Peat Swamp Forest appear at risk from fragmentation, and development may eventually reach their boundaries. The reserves cannot be managed in isolation from the surrounding land uses, some of which now threaten their hydrological balance and ecological integrity. The overall impact of logging within and outside the reserves has yet to be studied. The situation calls for a precautionary approach, integrating balanced multiple use with an emphasis on conservation. In this project, an attempt was made to integrate management through a consultative approach.

Malinau

The objective of the Malinau integrated spatial plan is to provide guidance for the development of Malinau District, taking into account the uniqueness and wealth of Malinau forests. The integrated natural resource management approach was implemented through participatory land evaluation and mapping processes.

Because the area is so remote, base map information was lacking. Data collected during the first phase of the project were insufficient. Intensive information gathering and surveys were undertaken, with some sampling reaching the village level.

Only a few villages were covered, but the effort so far can be used as a model for the local government.

Biodiversity

Pahang

The multidisciplinary assessment indicates the presence within the project area of individually distinctive forest subtypes besides peat swamp forest. Sungai Bebar, Merchong and their tributaries, together with their associated riverine vegetation, constitute the riverine habitats. A unique heath forest, characterized by sand ridges that represent ancient beaches, runs parallel to the coast. 'Islands' of dryland forest, such as Bukit Bangkung in Pekan Forest Reserve and several others in southern Resak Forest Reserve, are also inside the reserve.

Species restricted to peat swamps and specialist species, including those endemic to the Southeast Pahang region and the Malay Peninsula, have been recorded in the project site (UNDP/GEF, 2003). To date, 221of the 238 recorded floristic taxa can be found in the Southeast Pahang Peat Swamp Forest, among them *Gonystylus bancanus* (ramin), an important timber species of restricted distribution in Malaysia; *Durio carinatus* and *Tetramerista glabra*, food sources for birds and mammals; *Alstonia angustiloba*, which is a preferred nesting place for the globally vulnerable lesser adjutant stork; and *Nageia motleyi*, a relatively rare gymnosperm.

The Southeast Pahang Peat Swamp Forest is also home to diverse fauna, many of which are endemic. The forest supports 19 of the 33 IUCN 'red list' mammal species, 233 of the 237 recorded bird species (migratory species included) for Malaysian peat swamps, and 56 of the estimated 250 fish species recorded in Peninsular Malaysia.

Malinau

The first phase of the project in Malinau has resulted in better knowledge and understanding of the biodiversity richness of the area. The evaluation method developed during this phase has been widely implemented elsewhere (ITTO-CIFOR, 2002). A so-called multidisciplinary landscape assessment has confirmed the importance of local perceptions, landscape-level evaluation and biodiversity to local livelihoods.

The multidisciplinary landscape assessment, conducted during the first phase, revealed a wealth of information about Malinau's plants and wildlife. The flora survey recorded 173 families and 693 genera from 15,430 individuals collected. A total of 31 mammal species from 10 families were identified; five additional species were observed but not positively identified. Bird species were recorded in the survey area and the landscapes. Forty-seven fish species belonging to 33 genera, 13 families and three orders were recorded, and 13 insect orders comprising 79 families were collected.

In the second phase of the project, information derived from the first phase was shared with local people and officials. The detailed information has provided communities with a better understanding of the value of biodiversity and how it affects their livelihoods. The local designation of Malinau as a conservation district reflects that better perception of the important role of biodiversity. Through intensive communication and consultations with both communities and local officials, the project has facilitated an appreciation of the unique advantages of biodiversity.

Local Communities

Pahang

Living in the vicinity of the Southeast Pahang Peat Swamp Forest reserves are the Asli Jakuns, the indigenous inhabitants of Pahang. Numbering about 10,000, these people are one of the 18 subgroups of the Orang Asli of Peninsular Malaysia. They are dependent on forest produce for household, subsistence and dietary needs. They engage mainly in hunting, fishing and trapping of land and aquatic animals as well as collecting forest produce like rattan and medicinal plants.

Traditional Asli Jakun houses are constructed wholly using forest products: durian paya (*Durio carinatus*) timber is used for flooring, *merawan* (*Hopea* sp.) bark is used for walls, and *serdang* (*Pholidocarpus* sp.) or *palas* (*Licuala* sp.) leaves for roofing. The close interdependency of the Jakuns and the Southeast Pahang Peat Swamp Forest and its associated wetlands is a crucial factor for their survival.

Malinau

In East Kalimantan, Dayak people are the indigenous inhabitants of Malinau District. Several ethnic subgroups of Dayak live in remote and isolated areas and are connected with the outside only through rivers. A total of about 50,000 people of some 20 different ethnic groups inhabit the district. Dominant groups are the Kenyah, Merap, Lundayeh and Tidung. The majority of the people are swidden farmers, growing mainly rice. Cash income is obtained from supplementary activities, primarily the gathering of non-timber forest products. Access to information and decision-making processes has increased in the past eight years (Saloh et al, 2006). Communication between local communities and local officials has also increased through numerous meetings, training sessions, seminars and a workshop, all organized by the project.

Developing Management Plans

Strategy

Pahang

The development of the integrated management plans began in a systematic way. The conduct of this process was crucial and sometimes delicate since it addressed problems with 'business as usual', conflicts of interest, long-held attitudes and local capacity. During this process, stakeholders engaged in intensive dialogue using a participatory approach with the aim of harmonizing perceptions and accommodating different interests.

Data, maps, technical knowledge, syntheses and text drafts assisted in reconciling the views of various stakeholders. The participatory method was regarded as a way of harmonizing the bottom–up and top–down management approaches. The plans also had to meet the local legal requirements.

Malinau

Conveying information on spatial planning to all levels of government was a major achievement. The proposal for converting a large forest area to non-forest was put forward by the provincial government. The local government, uncertain about the expertise of its own staff, asked a working group facilitated by the project to review the proposal, consult with the parties and provide technical inputs to the decision-makers.

A presentation convinced the central government that the proposed forest conversion was technically unsound. The interest of one sector had dominated, and the plan did not conform to the existing land-use guidelines set by the national government. A simple biophysical evaluation showed how the proposed conversion ignored the importance of conservation and protected forest classification in favour of plantation and non-forest use. In response, the Minister of Forestry made reference to the importance of local and multilevel consultations, collaborative management and multidisciplinary knowledge.

The establishment of a multi-stakeholder working group on land-use planning in the District of Malinau is an example of policy change to implement integrated natural resource management.

Consultative planning

Pahang

It is important to address the conflicting interests of stakeholders in the management of natural resources. In Pahang, the issue was land-use allocation for agricultural areas and other development purposes versus the gazetted peat swamp forest reserves.

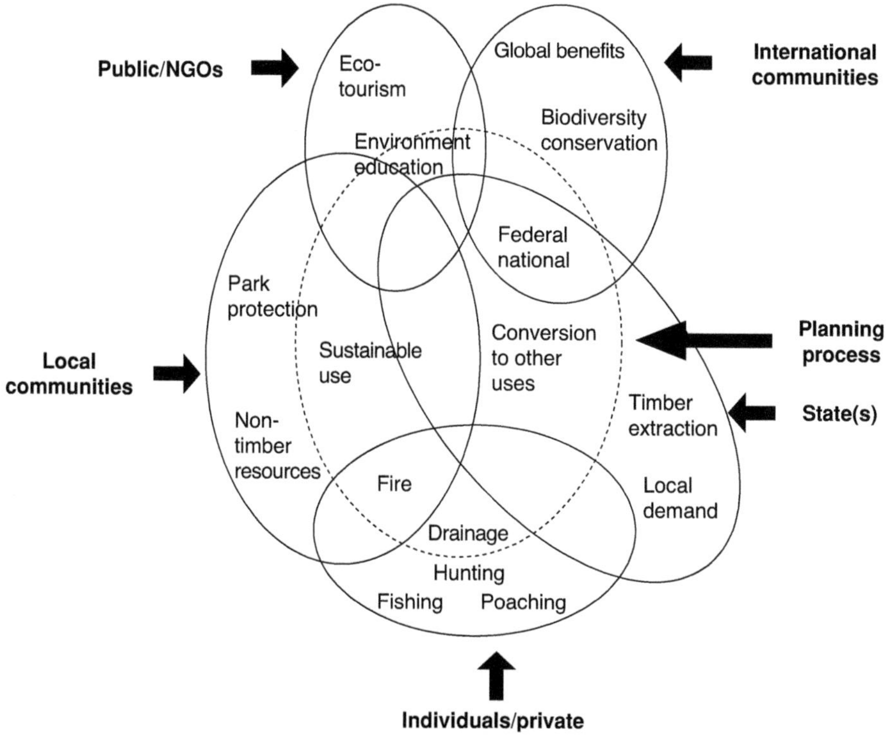

Figure 9.3 *Reconciling interests of multiple stakeholders in Pahang*

In this project, a new modus operandi allowed the establishment of a full planning process that was consultative (taking into account a range of stakeholders' views), cross-sectoral (involving relevant agencies), and multidisciplinary (incorporating knowledge on flora, fauna and socioeconomic features) (see Figure 9.3).

Addressing site-specific management issues presents an interesting management planning challenge; the project needs to facilitate the participatory process, adapting consultations with different stakeholders including local communities and the private sector in different groupings and stages.

Intensive development of the integrated management plan is being orchestrated at the local level. The expected plan will be an indispensable tool for planning sustainable forest management for the remaining unique peat swamp forest areas in Pahang. At the outset, the 'Five-S' framework (Nature Conservancy, 2003) was used to analyse the required planning elements: 'system, source, stress, strategy and monitoring success'.

The framework, which is site-specific, focuses on threat removal and conservation targets. With guidance from the specialists, the framework was adjusted to take site specification and actual local capacity into account. This framework encompasses ecosystem principles, multi-stakeholder participation and benefit sharing.

Malinau

Consultations in Malinau were undertaken with the district, province and central governments. The existing spatial plan was challenged by the district government because, with the new boundaries, the district has insufficient land allocated for development purposes. The question of finding a balance between conservation and development is a very sensitive issue in the district.

When Malinau declares itself a conservation district, it is expected that the central government will commit financial support to help the district while recognizing the environmental services provided by its forests. The amount of this support is still being discussed at the national level. The motive for the district to declare itself a conservation district has been the subject of a lengthy debate in national-level discussions and workshops. Malinau is not the only district to take this course, and others' intentions have attracted CIFOR and other national and provincial conservation organizations to facilitate discussions and workshops on the issue.

The motives range from demanding greater national support for conservation and seeking payment for environmental services to finding a better balance between development and conservation. At one extreme, advocates demand compensation for not exploiting the forest. At the other, the motive is said to be local wisdom in conserving the forest.

A Multi-tiered approach

Pahang

In developing the integrated management plan, a multi-tiered approach was adopted through two major stages of consultation: core planning and broader planning processes. The first stage began with the establishment of a management plan core team, which developed the draft integrated management plan with the help of a consultant who served as a facilitator.

The core team comprised representatives from 11 state government agencies: the Forestry Department, the State Economic Planning Unit, the Wildlife Department, the Department of Environment, the Department of Irrigation and Drainage, the Agriculture Department, the Department of Orang Asli Affairs, the Town and Country Planning Department, Pekan District and Land Office, the Land and Mines Department, and Pekan District Council.

Broader stakeholder consultations were held with other relevant agencies, the private sector (especially oil palm companies) and local communities. Consultations varied with the needs of individual stakeholder groups. An independent communications specialist and sociologist was engaged to assist in facilitating the consultations and to ensure that appropriate inputs from stakeholders were channelled to the integrated management plan. It is important to note that the consultations concerned the essential elements of the integrated management plan rather than the text of the plan per se.

During the consultation process, state authorities were kept informed about the discussions. Certain actions were agreed upon and implemented prior to the completion of the integrated management plan.

Crucial to the whole process was the leading role of the *Jabatan Perancang Bandar dan Desa* (Town and Planning Agency), which gave legal recognition to the integrated management plan by incorporating it into the Pekan District Local Plan (*Rancangan Tempatan Daerah*). As such, the integrated management plan will be fully recognized by all relevant state agencies as the primary reference in dealing with the Southeast Pahang Peat Swamp Forest (SEPPSF) and its associated management zones (*Kawasan Sensitif Alam Sekitar*) (Figure 9.4).

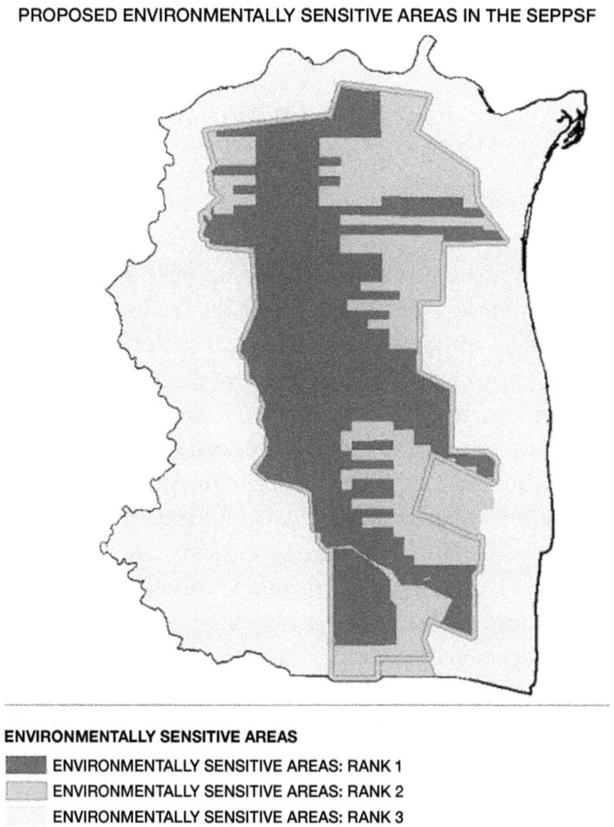

Figure 9.4 *Environmentally sensitive areas: Adopted concept in Pahang*

ESA Rank 1: No development, agriculture or logging permitted except for low-impact tourism.
ESA Rank 2: No development of agriculture. Sustainable logging and low-impact tourism may be permitted subject to local constraints.
ESA Rank 3: Controlled development where type and intensity of the development shall be strictly controlled depending on nature of constraints.

Malinau

In developing integrated natural resource management for Malinau, multi-level consultations were undertaken involving the government at the district, provincial and central levels. At the district level, the discussions took place in working groups that involved all land-related district agencies. Actively involved in this working group were the District Forest and Plantation Service, the District Agriculture Service, the District Mining Service, the Statistical Bureau and the Land Administration Service.

The core group in Malinau is a GIS (geographic information system) working group with representatives from different sectors. The role of the group is to facilitate and provide technical support. All results of the working group are presented to the local, provincial and central governments for their respective endorsement.

Planning Outcomes

Pahang

The Southeast Pahang Peat Swamp Forest Integrated Management Plan has been presented to various stakeholders, including the private sector, government agencies, local communities and NGOs. It is in the final stage of completion, with all its essential elements endorsed by the *Jawatan Kuasa Perancang Negeri* (State Planning Committee), and will be formalized by the State Executive Council.

Zoning, the principal strategy to reduce and control the identified threats in the Southeast Pahang Peat Swamp Forest, is a consensual process. Its aim is to maximize land-use benefits while maintaining landscape integrity. This determination of territorial demarcation is based on different physical environment profiles and varied land-use interests. Within the integrated management plan, zoning is a strategy to isolate and reduce potential threats to sustainability in the peat swamp forests. At the operational level, the demarcated zones will be recognized by stakeholders as management entities. Six management zones (see Figure 9.1) have been proposed:

- Zone 1: forest reserves for timber production and ecosystem protection;
- Zone 2: forest reserve extensions for the protection of ecosystem integrity through achieving connectivity;
- Zone 3: 1000m buffer zone for the reduction of impact of activities outside the forest reserves;
- Zone 4: 400m riverine buffer for the protection of associated wetlands;
- Zone 5: degraded and logged peat lands with the potential for development; and
- Zone 6: alienated and developed peat lands adjacent to peat swamp forest.

Each zone has a set of short- and long-term prescriptive actions leading to the achievement of the agreed objectives, including biodiversity conservation, maintenance of ecological integrity and ensuring sustainable management of the four forest reserves in Pahang.

The project has proposed to the state government that the entire Southeast Pahang Peat Swamp Forest be designated a *Kawasan Sensitif Alam Sekitar* ('environmentally sensitive area'), in line with the national physical plan concept (Figure 9.4). This designation has been incorporated into the Pekan District local plan. The proposal was accepted by the State Planning Committee in February 2006 and was endorsed by the State Executive Council, the highest authority in the state, in July 2006.

Malinau

The planning outcomes here are somewhat different from Malaysia's deliverables, perhaps in part because Malaysia has a long-established decentralized forest governance system. Given Indonesia's recent adoption of the system to decentralize forest governance to the district level, only the initial outcomes of intensive consultations and commitment building among stakeholders are apparent in Malinau.

Intensive consultations (Figure 9.5) with all stakeholders and at all levels of government have increased awareness of the important role of science-based

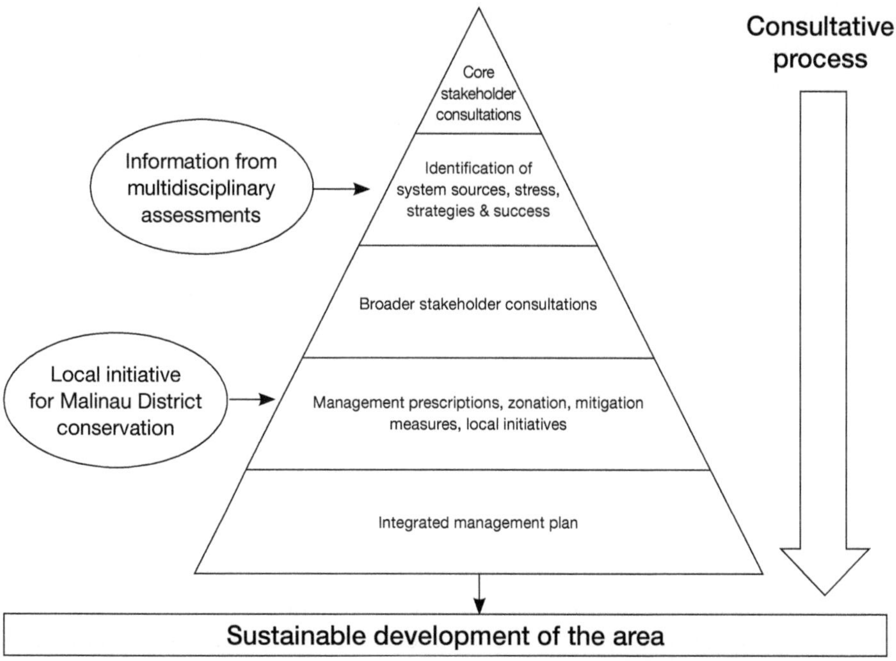

Figure 9.5 *Development of the management plan in Malinau*

land-use planning. The skills and expertise in the district have helped local government to make better use of reliable data and enhanced its ability to conduct land-use planning and engage multiple interests in determining the use of land resources in the district.

Conclusion

This chapter has highlighted initiatives taken in both areas, among other issues, to introduce a consultative planning aspect fundamental to the sustainability of the forest resources. This aspect was, to some extent, neglected in the past. However, other aspects involved within the governance context are rather complex and could not be addressed here.

The two projects are based on the assumption that the conservation and sustainable use of the respective areas' remaining forest must be addressed, taking the interests of various stakeholders as well as sustainability into account. The approach allows the establishment of a full planning process that is consultative, multidisciplinary and cross-sectoral, involving all levels of government and both public and private sectors. Although time consuming, this process is an effective way to reconcile the varying interests of stakeholders. Consensus is reached at the operational level, and a legitimate planning document can be written.

With the strong commitment of the state forestry department and support from the UNDP/GEF and other partners, the Pahang project managed to facilitate the development of an integrated management plan for peat swamp forests. This management plan is expected to be the basis for current and future decision-making on management issues pertinent to the forest reserves in Southeast Pahang.

In Indonesia, with the commitment of the central government and the support of the International Tropical Timber Organization (ITTO), the local government conducted appropriate planning processes for sustainable forest resource management, taking into account the biophysical features and other characteristics of Malinau District. The central government is expected to recognize and endorse the local initiatives that emerged from the stakeholder consultations in developing the conservation district of Malinau. This is expected to include incentives and financial assistance from the central government to support infrastructure development.

It appears from these two case studies that, in a decentralized forest governance system, mainstreaming a national commitment to conservation and sustainable forest management into local governance requires intensive stakeholder consultations and consultative planning of forest resources and, in turn, adoption of an integrated management plan at the local level. For relatively new decentralized governments, such as Malinau in East Kalimantan, local initiatives on conservation and integrated natural resource management required recognition and endorsement from the central government to be effectively implemented. Both cases demonstrate the importance of stakeholder consultation to achieve better understanding at the local level, to address the need for conservation and sustainability, and to serve as an effective tool in developing integrated management plans.

ACKNOWLEDGEMENTS

The authors acknowledge the assistance extended by the Forestry Department of Peninsular Malaysia, the Forest Research Institute of Malaysia, the Indonesian Forest Research and Development Agency, and the Center for International Forestry Research. Support from UNDP/GEF and the International Tropical Timber Organization for the respective projects is gratefully acknowledged.

REFERENCES

ITTO-CIFOR (2002) 'Technical Report Phase I 1997–2001. ITTO Project PD 12/97 Rev.1(F) Forest, Science and Sustainability: The Bulungan Model Forest', CIFOR, Bogor, Indonesia

Nature Conservancy (2003) *The Five-S Framework for Site Conservation: A Practitioner's Handbook for Site Conservation, Planning and Measuring Conservation Success*, Nature Conservancy, www.nature.org/summit/files/five_s_eng.pdf

Paramanathan, S. (2005) 'Soil resources of the Pekan-Nenasi Coastal Belt, Pekan District, Pahang Darul Makmur', UNDP/GEF funded Project MAL/99/G31 (unpublished)

Saloh, Y., Ford, A., Kristanty, B., Prianto, E., Achdiawan, R. and Ahmad, Z. (2006) '"Keberadaan CIFOR masih sangat diperlukan", hasil jajak pendapat masyarakat di Kabupaten Malinau' ['"CIFOR is still needed", the result of polling of communities in the District of Malinau'], internal CIFOR report, Bogor, Indonesia

Sheil, D. and Basuki, I. (2006) 'Informasi, dasar perencanaan penggunaan lahan yang baik mengalahkan perjudian dalam penggunaan lahan' ['Information, the important tool for land use planning beats gambling on land use'], *Warta Wanariset Malinau*, no X, August, CIFOR, Bogor, Indonesia

UNDP/GEF (2003) 'Southeast Pahang Peat Swamp Forest – Multidisciplinary assessment (MDA)', Final Reports, Volumes I and II, submitted by Wetlands International to UNDP/GEF-funded Project MAL/99/G31, 'Conservation and Sustainable Use of Tropical Peat Swamp Forests and Associated Wetland Ecosystems' (unpublished)

10

Implementing Decentralization: Lessons from Experiences in Indonesia

Kadjatmiko

In Indonesia, centralizing government authority used to be considered a way of preserving national unity, promoting economic growth and maintaining national stability. For more than 30 years, Indonesia enjoyed growth and political stability because authority was centralized. Economic growth was based on centralized authority even though Indonesia is an archipelagic country comprising various cultures, ethnic minorities and diverse communities. Highly centralized public sector decision-making, however, reduces accountability, retards the development of public infrastructure, adversely affects rates of return on public sector projects and constrains the development of local institutions.

Between 1945 and 2001, during the era of centralized government, at least three systems of financial transfers were in operation. First, from 1945 to 1956, the national government provided government subsidies to ensure local governments' ability to finance their planned budgets. By 1956 the national government had issued Law No 32/1956, with the concept of an intergovernmental transfer formula. The government introduced revenue sharing so that local governments received a percentage of government tax revenues. In 1972 the so-called subsidy for autonomous regions (SAR) was implemented, which implied that salaries of local government civil servants were covered by the national government. Over 60 per cent of the SAR was allocated to provincial governments, with the remainder allocated to the districts. Local governments also received President's Instruction transfers, which covered local development activities.

Before 1998, Indonesia was probably considered the most centralized country in the world, in that the authority to determine most aspects of national policy was held by central government. The revenue and expenditure systems were fairly centralized, and central government collected almost 95 per cent of national revenue, allocating 87 per cent of that for expenditure and transferring the remaining 13 per cent to local governments as grants and subsidies. National revenue was derived from taxes (59.2 per cent), national resources (23 per cent) and aid (17.8 per cent), and funds to provincial governments, at the first level, came from a tax share, general purpose grants (GPGs) and development grants. Revenue for regency governments, as the second level, came from first-level (provincial) government, local taxes, routine expenditure transfers (under the SAR) and development expenditure transfers as President's Instructions. Central government fully supported the costs of local government employees under the SAR scheme. For this reason, the SAR is classified as a specific purpose or conditional grant, with no discretion at the local government level, to support routine activities.

To support local development activities, the central government allocated President's Instruction funds, which have the important objective of reducing regional disparities. These are transferred as specific purpose grants (SPGs) as well as SAR (for example for basic education) and as block grants (for example for provincial development).

In the Reform (*Reformasi*) Era, beginning in 1998, Indonesia implemented a policy of regional autonomy. The policy was embodied in Laws No 22/1999 and No 25/1999 and fully implemented in 2001. These two laws provide a new framework for a relationship between local governments (districts and provinces) and the central government. These laws devolved all authority to the local governments except for authority for national defence, monetary affairs, justice, development planning and religion.

Implementing regulations were passed to support Law No 33/2004 as an amendment of Law No 25/1999. In accordance with this amendment, Government Regulation No 104/2000 was revised as Government Regulation No 55/2005 on fiscal balancing. This regulation was intended to devolve functions to the regions. Such a design works well only if the central government budget is healthy, however, and unfortunately this is not the case in Indonesia. With a huge national debt, coupled with economic and social crises, funding for decentralization was limited.

By providing only very limited funding, the central government expects that local governments will eventually adjust their expenditures accordingly. The 'money follows functions' rule is not being applied by the central government in the distribution of funding. Instead, it may be inferred from the budgetary situation that it is more a case of 'functions follow money', especially at the local (district and municipality) level. With limited local funding and transfers, it is expected that local governments will not be able to support all the functions delegated to them. This problem is coupled with the fact that the functions assigned to local governments have never been clearly defined. It is better, therefore, to refer to this local situation as 'functions (only) follow (the availability of) money'.

Two laws on special autonomy were also enacted in 2001. These two laws – the Law on Special Autonomy for Aceh and the Law on Special Autonomy for Papua

– were passed as a result of local aspirations and the rather atypical needs to solve the problems of vertical imbalance. The laws made 'the asymmetric treatment' of sub-national government legally acceptable. That is, Papua was given 70 per cent of revenue-sharing rates for oil and gas and also receives special grants equal to 2 per cent of the national GPG transfer. Both Aceh and Papua provinces receive 15 per cent of oil revenues and 30 per cent of natural gas revenues plus an additional 55 per cent of oil revenues and 40 per cent of natural gas revenues because of their special autonomous status (total shares for both provinces being 70 per cent: 15 + 55 per cent and 30 + 40 per cent of oil and natural gas revenues respectively); these funds are included in the state budget as shared revenue, not as funds for special autonomy.

Decentralization began in Indonesia after more than three decades of a centralized system. Indonesia's large population and vast archipelago, along with the aspirations of districts and provinces, required a new, more decentralized system. Decentralization began, in fact, with a spirited declaration by the *Reformasi* movement without the benefit of adequate time for planning and preparation.

The decentralization of government in Indonesia was conducted quickly, by devolving to local government the authority for everything except the five functions mentioned earlier – functions considered to have nationwide impact. Decentralization, in turn, encourages the regions to establish more local governments to deliver better public services. As shown in Table 10.1, as a result of decentralization, the number of provinces has grown by 10 per cent, and the number of districts or cities has grown by more than 25 per cent.

Assignment of Expenditure

The national government has responsibility for national public services, defence, security, international affairs, monetary policy, legislation, transfers and subsidies to persons and businesses, fiscal policy coordination, regional equity, redistribution, and the preservation of internal common markets; local governments have authority and oversight devolved to them as described above. Local services are assigned to local governments. Where responsibilities are shared, the role of each level of government must be clarified.

The economic issues involved in the financing of provincial and local governments deserve and demand separate attention for three reasons:

1 the provincial and local governments are now a substantial part of Indonesia's economy, accounting for approximately 5.94 per cent of gross domestic product and representing about one third of total consolidated government expenditures;
2 the major services provided by government that most affect citizens on a day-to-day basis – ranging from basic education, health and transportation to social

Table 10.1 *Percentage share of state revenues before decentralization and after implementation of Law No 33 of 2004*

No	*Revenue type*	*Before decentralization*			*After decentralization*			
		Central	*Provinces*	*Districts*	*Central*	*Provinces*	*Districts*	*Balance to other Districts*
1.	Property Tax	10	16.2	64.8	–	16.2	64.8(+)	+
2.	Property Rights Transfer Tax	20	16	64	–	16	64(+)	+
3.	Personal Income Tax	100	–	–	80	8	–	–
4.	Forestry: IHH	55	30	15	20	16	64	–
5.	Forestry: PSDH	55	30	15	20	16	32	32
6.	General Mining: Land rent	20	16	64	20	16	64	–
7.	General Mining: Royalties	20	16	64	20	16	32	32
8.	Fishery	100	–	–	20	–	–	80
9.	Oil	100	–	–	85	3	6	6
10.	Gas	100	–	–	70	6	12	12
11.	Reforestation Fund	100	–	–	60	–	40	–
12.	Geothermal	100	–	–	20	16	32	32

Source: Law No 33 of 2004

services and public safety – are dominated by provincial and local governments; and

3 because of the diversity of provincial and local governments and the magnitude and mix of revenues and expenditures, many economic and fiscal issues are substantially different for provincial and local governments than for the central government.

Under the expenditure approach, functions are first designated as the clear responsibility of one level of government. Thus functions such as primary education, local general hospitals, clean water and refuse collection would be assigned to local governments as having primarily local impacts and being within the local management capacity. Specialist secondary schools, regional specialist hospitals, and river and water resources management would be assigned to the provincial level as having substantial inter-regional impacts. Religious affairs and justice would be reserved as central government functions, requiring uniform policies with nationwide impact. Once the assignment of functions is agreed upon, the allocations of financing sources between levels and among the regional governments at each level are tailored accordingly.

Assignment of Revenue

According to Law No 25/1999, local governments (provincial and district or municipal) are encouraged to increase their own revenues from various sources, such as local taxes and charges, profits from locally owned enterprises, and other eligible local revenues. However, the current scheme still provides only a small proportion of local revenue compared with total national revenue. The central government still controls the major revenue from important sources – income taxes, value-added taxes and others – whereas local governments have only weak local taxation power.

Intergovernmental Transfers

Districts and cities have essentially become responsible for all public services that the central and provincial governments do not deliver in these areas: public works, health, education and culture, communications, industry and trade, capital investment, environment, land, and cooperatives and labour. A third of the national budgetary expenditure is transferred to local budgets, mostly as block grants. Almost 26 per cent of national revenues are transferred to regions through the general purpose grant mechanism. The management of 2 million national civil servants has also been transferred to the regions.

After Laws No 22/1999 and No 25/1999 were enacted, decentralization evolved gradually. Consequently, as a result of widespread calls for clearer decentralization and popular pressure to maintain the pace of development through more rapid growth of services, the government and parliament revised Laws No 22/1999 and No 25/1999 by enacting Law No 32/2004 on Local Government and Law No 33/2004 on Fiscal Balancing between Central Governmental and the Regions.

The amendments to Laws No 32 and No 33 of 2004 were intended to improve the implementation of decentralization. Indonesia's fiscal decentralization is supposed to increase the efficiency of national allocation by devolving authority from central government to the sub-national level. The devolution of authority was also intended to meet regional aspirations, improve the fiscal structure, and mobilize regional and national revenues. Laws No 32/2004 and No 33/2004 are intended to meet the 'finance follows function' principle. Based on this principle, the balancing fund is designed to ensure that each local government can deliver mandatory public services using its fiscal capacity. Since the Indonesian decentralization process is intended to ensure equity, there should be fair treatment for each local government to close the gap between needs and capacity. It is very important to ensure that all Indonesians are provided with the same minimum basic services by their local governments. However, allocating the same amount of money to each local government will not mean fair treatment, since local governments have a variety of different needs and capacities. The allocation has to be designed so that every local government can at least bridge the gap between its minimum needs and its maximum fiscal capacity.

Under Law No 33/2004, local governments control important revenue resources (taxes on hotels and restaurants, royalties, entertainment, advertising, parking, street lighting and selected mining), the provinces control more important resources (motor vehicle registration, transfer and fuel taxes and water-user fees), and the central government manages the most important and productive sources (local personal income, property rights transfers and other shared revenues). According to the law, regions receive decentralization funds, deconcentration funds and co-administered tasks. 'Decentralization' means the transfer of the authority of the government to an autonomous region to regulate and administer governmental affairs in the region under the Unitary State of the Republic of Indonesia. The transfer is followed by a fund transfer, called the decentralization fund, to regional budgets. 'Deconcentration' means the delegation of governmental authority to the provincial governments. 'Co-administered tasks' refer to assignments from the government to a region and/or village or other entities with the obligation to report and account for the implementation of the assignment to the government. Deconcentrated and co-administered tasks are funded by the national budget through line ministries.

Intergovernmental fiscal transfers play a critical role in public finance in both unitary and federal decentralized countries. In the case of decentralization reforms, the design and implementation of transfers have significant impacts on the potential efficiency and equity of basic public service provision. Indonesia's 2001 decentralization process illustrates the implementation of significant reforms in changing to an intergovernmental fiscal system. Almost overnight, local governments were effectively put in charge of almost a third of consolidated government expenditures (Hofman et al, 2006), whereas their 'regional own revenues' (*Pendapatan Asli Daerah*, PAD) were able to finance only approximately 7 per cent of their expenditures. To deliver public services closer to the point of use, local government tends to expand. At present, local government sources of revenues are PAD, balancing funds and other income. Since the enactment of Law No 25/1999 local governments have had the power to raise local taxes and user fees, although under devolution they can raise local taxes only up to 3.5 per cent of national taxes. Balancing funds are derived from revenue sharing and general purpose and specific purpose grant transfers.

Revenue sharing is intended to respond to regional aspirations for increased access to and control over revenues, strengthen regional ability to mobilize revenue, and address vertical fiscal imbalance. Revenues consist of taxes and 'non-taxes' (income from natural resources). According to Law No 25/1999, the revenue is shared according to origin – that is, by producer region. The balancing funds transfer system is intended to address vertical imbalances between levels of government (revenue sharing and general purpose grants), equalize local government fiscal capacities, encourage local expenditure on national development priorities, promote the attainment of minimum services and standards, and stimulate revenue mobilization.

Revenue Sharing

Revenue sharing is intended to respond to local pressures, stimulate local participation in national revenue mobilization and address vertical fiscal imbalance. Sources of revenue to be shared include natural resources (forestry, general mining, fisheries, oil, gas and geothermal energy) and taxes (property taxes, property rights transfer taxes and personal income taxes).

Nevertheless, many regions remain unhappy with the revenue-sharing mechanism and believe that they do not receive fair shares. Revenue sharing is also intended to alleviate vertical imbalance, thereby reducing local government budget deficits; however, revenue-sharing systems potentially create horizontal imbalance between regions, depending on their natural resource wealth and other factors.

Sharing of natural resource revenues is considered an inappropriate instrument for fiscal decentralization because natural resources are usually unevenly distributed, and this can induce higher horizontal fiscal imbalances across regions. Of course, natural resource revenue sharing may be justified as a tool to improve vertical fiscal imbalances. However, in general, its negative impact on equitable revenue allocations is much bigger than its positive effect on the vertical fiscal imbalances. It is for this reason that sharing of natural resource revenues is not adopted in many countries (Sidik, 2003).

General purpose grants

According to Law No 33/2004, general purpose grants (GPGs) should act as an equalization measure. Furthermore, GPGs should be sufficient to close the gap between minimum needs and maximum possible fiscal capacity. Consequently, each local government receives a different level of allocation, according to its needs and capacities. The variation in needs may arise from differences in the size of the population or area, geography and people's welfare; the variation in fiscal capacities may come from differences in tax revenue, non-tax revenue and PAD. The GPG is also used to reduce horizontal imbalances resulting from revenue sharing across regions. It addresses vertical fiscal imbalance between levels of government. The grant equalizes regional government fiscal capacities in order to deliver services and stimulate revenue mobilization. It is also intended to respond to regional aspirations for greater access to and control over revenues. Moreover, the amount of the grant is determined by fiscal gap analysis – determining fiscal needs minus fiscal capacity – which means that the formula recognizes the fiscal capacities (capabilities) of regional governments to finance their expenditures. Regions with lower fiscal capacity receive more in GPG, and vice versa.

Districts and cities are allocated 90 per cent of GPG funding, and the remaining 10 per cent is allocated to the provinces. These grants are intended to address problems related to both vertical and horizontal fiscal imbalances. The formula adapts the equalization principle and considers measures of both expenditure needs and fiscal capacities. Expenditure needs are derived from an analysis of population size, area, consumer price index, human development index and gross regional domestic per capita product; the fiscal capacity is derived from PAD and revenue sharing. The grants account for an average 64 per cent of total district revenues.

The GPG acts like an equalization mechanism, but there is a balancing factor, known as 'hold harmless'. This means that local governments that received large transfers in the past will not receive less. GPGs were first allocated in 2001 (see Table 10.2); only 20 per cent of the grant was then allocated, by means of a formula, and this was based on the 'hold harmless' principle in relation to the previous year's subsidy for autonomous regions and President's Instruction transfers. In 2002 the formula-based allocation increased to 40 per cent; the remainder was allocated on the basis of lump-sum 'hold harmless' factors. The lump sum implies that a minimum amount was equally distributed across regions. Only 10 per cent of total GPG was allocated through the lump-sum mechanism in 2002. Fifty per cent was allocated through the 'hold harmless' policy, which meant that no region received less than in 2001. In 2003, 2004 and 2005 the GPG was essentially allocated along similar principles, although the formula-based share was increased and figures were updated to account for the increase in the number of regions. New 'hold harmless' conditions have superseded the 2001 'hold harmless' conditions. Simulations were always made using the previous year's local government figures.

Table 10.2 *General purpose grants (Rp trillion)*

Year	*Total*	*Provinces*	*Districts/Cities*
2001	63,609.00	30	336
2002	71,168.87	30	348
2003	79,240.40	30	370
2004	83,139.37	32	410
2005	89,571.10	32	434

Source: State Budget legislation 2001–2005

Indonesia will implement the 'hold harmless' concept until 2007, such that no local government receives a lower grant than it received in 2005 (according to the GPG formulation, regions may receive a lower or higher grant than in the previous year; however, equalization funds are allocated to regions receiving lower grants). The formula used to calculate GPGs is given in the Annex on pages 161–162.

Specific Purpose Grants

Initially, the specific purpose grant (SPG) had two components: the reforestation specific purpose grant, which was a mechanism for intergovernmental sharing of proceeds from forest exploitation, and the regular specific purpose grant. The reforestation SPG was first allocated in 2001, and both SPGs were allocated in 2003 (see Table 10.3). By 2006 the reforestation specific fund had become a revenue-sharing mechanism, which meant that in 2006 the allocation of SPGs was for non-reforestation purposes only.

The reforestation SPG is allocated on the basis of the 40 per cent share of reforestation funds that were allocated to regions at the provincial level. The SPG is allocated annually according to the national budget to help certain regions finance special activities that are purely regional affairs. The government establishes several criteria when allocating SPGs: general criteria, special criteria and technical criteria.

The general criteria are established with due regard to the financial capacity of the region, and the special criteria apply to the prevailing laws, regulations and characteristics of the region. Technical criteria are established by the line ministries. A region receiving an SPG must provide matching funds of at least 10 per cent to establish a regional commitment to the investment.

Special criteria are established by using dummy variables to assess whether regions meet the criteria and conditions as stated in the decentralization law.

In 2003 specific activities funded by non-reforestation SPGs were primary health, primary education, infrastructure and local government. In 2006 these expanded to include primary health, primary education, infrastructure (roads,

Table 10.3 *Specific purpose grant (SPG) allocation (Rp billion)*

Sector	*2003*	*2004*	*2005*	*2006*
Primary Education	625.00	652.60	1221.00	2919.53
Primary Health	375.00	456.18	620.00	2406.80
Roads	842.50	839.05	945.00	2575.71
Irrigation	338.50	357.20	384.50	627.68
Clean Water	0.00	0.00	203.50	608.00
Local Government	88.00	228.00	148.00	448.68
Fishery	0.00	305.47	322.00	775.68
Agriculture	0.00	0.00	170.00	1094.88
Environment	0.00	0.00	0.00	112.88
Total	2269.00	2838.50	4014.00	11,569.80

Source: State Budget legislation 2001–2005

bridges, irrigation and clean water), fisheries infrastructure, agriculture, local government infrastructure and the environment.

Of the total government expenditure of Rp 647.7 trillion, SPG 2006 transfers were Rp 11.57 trillion, or only 5.2 per cent of total local expenditure (transferred funds) or 1.7 per cent of government expenditure. Yet SPGs remain an important mechanism for maintaining the delivery of public services. The formula used to calculate SPGs is given in the Annex on pages 161–162.

Problems and Challenges

The mismatch between grants transferred and functions devolved to the regions is caused mainly by the rather simplified transfer formulas, which do not accommodate complicated functions. The unclear responsibilities of the devolved functions has also tended to create overlapping funding, that is through the fiscal balancing mechanism (intergovernmental fiscal transfers) or through the line ministries fund (deconcentration fund). The contribution of PAD to local revenue was low because most local governments had been assigned only limited devolved revenue. Local governments mistakenly assumed that central government would pay local civil servants because local civil servants' salaries were computed in the GPG formulation. This, of course, gave them an incentive to increase spending on personnel at the cost of other priorities.

The devolution of authority was followed by the devolution of expenditure responsibilities, as regulated in the decentralization laws. Although they still deliver national public service priorities, local governments may not be able to afford such responsibilities because of the limited funds they receive, both shared and through GPGs. The only possible mechanism is the SPG, which covers national priorities. However, some in central government believe that it is not right to achieve national objectives through this mechanism if it could be achieved through the deconcentration mechanism. The issue of whether to provide financing through specific grants or through the deconcentration mechanism has been widely disputed. Government Regulation No 55/2005 implied that the deconcentration fund should be gradually transferred to the SPG; however, the dispute persists because line ministries still see some local functions as national functions.

The decentralization legislation devolves authority for governance for all issues except international policy, defence and security, judicial and monetary affairs, and religion. Legislation relating to line ministries, however, still covers aspects of governance not included in those exceptions. Line ministries continue to control significant line expenditures (for example for capital and development) in devolved sectors, such as basic education, health and infrastructure. At least 25 laws regulating line ministries contradict the laws on decentralization. Law No 33/2004, for example, implies that deconcentration (line ministries) and jointly administered tasks that are funded as part of the line ministries' budgets and utilized to administer regions will gradually be transferred to SPGs, but the prevailing law relating to line ministries states the opposite.

Law No 33/2004 implies that the 'hold harmless' policy for allocating GPGs will be discontinued beginning in 2008, so the possibility exists that some regions will receive less GPG than in 2005 – or none at all. The idea of this regulation is to make the allocation formula more just and fair.

Fluctuation in national revenues will in turn affect the transfer of funding to the regions; that is, fluctuation in national revenues will result in fluctuation in transfers. This must encourage both central and local governments to increase their revenues.

There are issues of transparency in allocating the transfers. The law on decentralization implies that there are formulas for allocating GPGs and SPGs, along the lines of the revenue-sharing systems, but protests continue about the level of funds. Many still believe that the formulas are being manipulated through lobbying at the central government level and in the legislature. Furthermore, the regions do not consider the GPGs and SPGs to be fairly allocated.

Differences between national and local objectives create disputes over SPG allocations in particular. To receive their allocations, the regions are obliged to utilize the funds according to national objectives; however, some regions maintain that their local objectives do not match the national objectives. These differing objectives could mean that the funds transferred might not be available for use. For example, a region may receive an SPG to finance the primary education sector, with the guideline that the funds must be used to rehabilitate classrooms, while the local government actually may need to build more schools; consequently the money could not be used and there would be no point in making this transfer.

The GPGs allocated at least 25.5 per cent of net domestic revenue, and this will increase to 26 per cent by 2007. The regions contend that this percentage is far too small compared with their total needs. Although local governments are encouraged to increase their PADs, the Association of Local Governments recommends that the transfer to local governments be at least 40 per cent to fund the devolved functions.

Clearly, decentralization has brought complexities and confusion for Indonesia, along with the desired greater local autonomy.

References

The following list includes publications referred to in preparing this chapter, as well as those cited in the text.

Alm, J., Martinez-Vazquez, J. and Indrawati, S. M. (2004) *Reforming Intergovernmental Fiscal Relations and the Rebuilding of Indonesia: The 'Big Bang' Program and its Economic Consequences*, Edward Elgar, Cheltenham, UK

Directorate General of Fiscal Balance (2004) *Intergovernmental Fiscal Review 2001–2003*, Directorate General of Fiscal Balance, Jakarta

Hofman, B., Kadjatmiko, Kaiser, K. and Sjahrir, B. S. (2006) *Evaluating Fiscal Equalization in Indonesia*, World Bank, Jakarta

Ministry of Finance (2000–2005) *National Budget 2000–2005*, Ministry of Finance, Republic of Indonesia, Jakarta

Ministry of Finance (2002) *Fiscal Decentralization: Ministry of Finance Policy Agenda*, Ministry of Finance, Republic of Indonesia, Jakarta

Ministry of Finance (2002) *General Purposes Grant, Concept, Burden and its Prospect*, Ministry of Finance, Republic of Indonesia, Jakarta

Morris, N. (2002) *Fiscal Capacity and Expenditures at Different Levels of Government*, Asian Development Bank, Melbourne, Australia

Sidik, M. (2003) *Indonesia's Imbalance: Decentralization and its Future Direction for a Greater Taxation Power to Sub-National Governments*, Ministry of Finance, Republic of Indonesia, Jakarta

Sidik, M. (2005) *East Asia Decentralizes: Making Local Government Work*, World Bank, Jakarta

Sidik, M. and Kadjatmiko (2002) *Indonesia's Fiscal Decentralization: Combining Expenditure Assignment and Revenue Assignment*, Ministry of Finance, Republic of Indonesia, Jakarta

ANNEX: FORMULAS USED TO CALCULATE GPGs AND SPGs

GPGs

GPG = total wages of local civil servants + fiscal gap (FG) (1)*
FG = fiscal needs – fiscal capacity (2)

Fiscal needs (FN) are derived as follows:

$$FN = Abe\ (\alpha_1 IP + \alpha_2 IAr + \alpha_3 CPI + \alpha_4 HDI + \alpha_5 GRDP) \quad (3)$$

where:

Abe = average of total local budget expenditure
IP = population index
IAr = area index
CPI = construction price index
HDI = human development index
GRDP = gross regional domestic product
α_1, α_2, etc. = coefficients.

The coefficients α_1, α_2, α_3, α_4, and α_5 in Equation 3 are distributed by 'best equalization', according to the Coefficient of Variation utilizing the Williamson Indexation.

Fiscal capacity (FC) is derived as follows:

$$FC = PAD + RSNa + RST \quad (4)$$

where:

PAD = 'regional own revenue'
RSNa = revenue sharing of natural resources
RST = revenue sharing of taxes.

SPGs

General criteria represent regional fiscal capability and imply regional capacity to cover the development expenditure. Thus the SPG is meant to finance the development of the regions, especially those sectors relating to national objectives.

$$\text{RFC} = \text{general revenue from local budget} - \text{civil service expenditure} \quad (5)^*$$

and

$$\text{General revenue} = \text{PAD} + \text{GPG} + \text{RS} \quad (6)$$

where:

RFC = regional fiscal capability
PAD = 'regional own revenue'
RS = revenue sharing.

Note: *Law No 33/2004.

11

Devolved and Decentralized Forest Management in the Philippines: Triggers and Constraints in Investments

Ernesto S. Guiang, Ferdinand Esguerra and Domingo Bacalla

This chapter reviews the history of devolved and decentralized forest management in the Philippines, the resulting policies and their impacts. Building on previous reviews on related topics by Brillantes (2000), De los Angeles (2000), Mercado (2000), Borlagdan et al (2001), DAI (2002), Magno (2003), the World Bank (2003 and 2005b), Bacalla (2006) and Pulhin (2006), it analyses how these policies, coupled with emerging opportunities and the current public agenda, either facilitate or hinder investments in devolved and decentralized forest management and thus shape the future of Philippine forestry.

In *decentralization*, central government functions are transferred to the regions but officials and functions remain under the control of the same department, office or ministry; under *devolution*, the national government confers powers and authority to local government units to perform specific functions and responsibilities (Section 17(e), Republic Act 7160, otherwise known as the Local Government Code).

We focus here on the widening gap between the written policies of the national government and the reality at the local government and community levels and document the growing 'disconnect' between what was intended and what is actually happening in forest management in the Philippines. We then offer recommendations based on our reflections, field observations and institutional memory, and suggest a framework for triggering improved decentralization and devolution that balances theory, needs for incentives, practicality and required support systems appropriate for the Philippine environment.

Historical Background

Before and during the Marcos regime (from the post-war years to 1985), forest management planning, allocation and regulation were largely centralized and heavily influenced by national politicians and vested interest groups. Local government units and communities, including indigenous people, had minimal influence on how the forests were managed. State and elite capture of the high-value natural forests for exploitation dominated the sector and the industry (Porter and Ganapin, 1988; Vitug, 1993; Coronel et al, 2004).

The end of the Marcos regime in 1986 through the 'People Power' revolution paved the way for a major shift in forest management. As citizens clamoured for more participation in governance, opportunities opened up for decentralized and devolved forest management in the Philippines. By the early 1990s, most of the timber licence agreements, mainly controlled by friends of the previous administration, had been cancelled, suspended or not renewed. When the licence-holders pulled out, their areas of operation – covering thousands of hectares of forest land – became 'open access' and attracted migrants who, seeing the forests as a good and a (in many cases the only) source of livelihood, engaged in illegal logging, forest conversion, slash-and-burn farming and other destructive practices. The 1987 Constitution further limited the entry of the private sector into the forest industry by allowing them to gradually convert their timber licence agreements (TLAs) into joint ventures, co-management and co-production agreements.

Despite the politically fragile situation, the donor community and pro-democratic institutions responded positively to the leadership change. A master plan for forestry development that promoted conservation, production, equity and alleviation of poverty was crafted in 1990. As a result, close to US$1 billion from the donor community and the government of the Philippines (counterpart financing) was allocated for the sector to support decentralized, devolved and community-based initiatives in forest management. 'Equitable access to natural resources' and 'encouraging participatory processes' at the local level became the call of the day. At the time of writing, however, no records are available on how much of the foreign and local investments really went to decentralized and devolved forest management.

The advocacy for more devolution resulted in the passage, enactment and issuance of various policy instruments. In the forestry sector, these policy changes started to chip away at the power of the Department of Environment and Natural Resources (DENR) and the forestry bureau to reclassify lands, issue resource-use rights and permits, title publicly released alienable and disposable lands, and issue environmental compliance certificates (Guiang and Castillo, 2005; FMB, 2006). These policies led to a multimodal and institutionalized decentralization and devolution of forest management in the Philippines (Magno, 2003). The process has always been an uphill struggle.

Aside from the 1987 Philippine Constitution, which provided for a system of decentralization, other significant laws and policies were promulgated to push decentralization and devolution (Box 11.1).

Box 11.1 Laws and policies encouraging decentralization and devolution

- Presidential Executive Order No 192 (1987), which reorganized the DENR;
- the Local Government Code of 1991, which provided for partial devolution of forest management functions to the local government;
- the National Integrated Protected Area System of 1991, which provided the mechanism for local stakeholders to participate in managing protected areas through the creation of local boards;
- Executive Order No 263 (1995), which mandated the DENR to adopt community-based forest management as the national strategy;
- the Indigenous Peoples' Rights Act of 1997, which transferred from the DENR to the National Commission on Indigenous Peoples the power to process legitimate claims to ancestral domains, further strengthening recognition of indigenous peoples' right to manage the resources covered by their ancestral domains; and
- the Philippine Clean Water Act of 2004, which mandated the DENR to organize water resources boards composed of local officials and representatives of NGOs and national government agencies.

These policies were considered adequate, well formed and pioneering. Implementing guidelines were then formulated, providing clearer definition of concepts and ideas as well as laying down standards, systems and processes to make decentralized or devolved forest management a reality at the local level. But financing and investments in decentralized and devolved forest management, while obviously considered during the stage of planning and formulation of policies, were insufficient: resource allocations and investments did not meet even the minimum requirements for translating policy into action.

Types of Decentralized and Devolved Forest Management

At least six types of decentralized and devolved forest management have emerged since 1986 (Borlagdan et al, 2001; Contreras, 2003), as described in Box 11.2.

A closer look at each type reveals an ongoing power struggle between the DENR, non-DENR national agencies, local governments and communities.

Box 11.2 Types of decentralized or devolved management in the Philippines

- forests devolved to upland communities, such as those covered by community-based forest management agreements;
- watersheds decentralized to DENR field offices;
- protected areas decentralized to local protected area management boards and water quality management area boards (both of which include local stakeholders);
- forest lands devolved to local governments, such as those under co-management agreements;
- forests devolved to non-DENR national agencies, such as those managed by the National Irrigation Agency, the National Power Corporation, the Philippine National Oil Company and academic institutions (like the University of the Philippines in Los Banos); and
- forests devolved to indigenous peoples, such as those covered by certificates of ancestral domain title or certificates of ancestral land title.

Except for those lands devolved to the indigenous peoples and to the non-DENR national agencies, all decentralized and devolved forest management in the Philippines remains under some kind of DENR supervision, which has been highly criticized by the local governments (Agbayani, 2005). The DENR continues to be the major player in devolved community-based forest management areas, multi-stakeholder and inter-agency oversight of protected areas, water quality management areas, watershed reservations and co-managed forest lands. Although the local governments, expected to provide financial and manpower support, now serve as the balancing factors in co-managed areas and protected areas, the DENR has retained the power to issue (and suspend) resource-use rights and environmental compliance certificates in all these areas. This suggests that the DENR is not ready to devolve other functions to the local governments, which in turn appear not prepared to carry out the devolved and decentralized forms of participatory forest management that are now within their powers.

For example, some local government officials who sit as co-chairs of protected area boards appear unwilling to assert their authority in implementing or enforcing board resolutions. In these areas, devolution is taking the form of 'partnership and collaboration' with the DENR; the local governments are not taking the lead.

In an ideal setting, decentralized and devolved forest management should provide the proper mix of responsibility among local and DENR officials and other local stakeholders (with transparency and accountability) together with incentives, benefits, consistent policies, financing, safety nets and explicit authority. Balancing the above concerns has become difficult, however, because forest management requires a 'social enterprise' approach (Dacanay, 2004). Depending on the mix of objectives, forests have to be managed to produce multiple forest goods and services, including biodiversity, social justice and profit.

Incentives and Constraints

Communities, local governments and the DENR respond to different incentives for decentralized and devolved forest management. Incentives for communities to adopt sound forest management practices range from security of tenure to awarding of resource-use rights plus others that may be allowed by the DENR. Guiang and Castillo (2005) argued that, compared with other tenure holders, holders of community-based management agreements in forest lands, watershed reservations and protected areas enjoy the fewest rights under their tenure instruments. Community incentives under the different types of decentralized and devolved system largely depend on stable policy and sustained funding support for forest management. They may include employment, access to technology and support, social infrastructure, and linkages to the private sector.

For the DENR to champion decentralized or devolved forest management, the incentives are mainly for professional growth, networking, career paths and good relations with local governments and, thus, local allies in enforcing rules and regulations, resulting in reduced expenses for forest protection and overall management.

The sudden surge of decentralized and devolved policies in forest management has not, however, provided adequate time for the DENR, local governments and communities to develop and implement adequate standards, systems and processes for managing forest lands. The forestry sector, for instance, is still adjusting to the change from timber concessionaires to community-based forest management.

The tenure holders, private interests and local governments also operate under many constraints that raise the cost of doing business in the forestry sector (see Box 11.3).

To implement decentralized and devolved forest management, concerned institutions and resource managers have to develop and install standards, systems and processes that reduce the overall cost of doing business. Currently, there is no unified set of standards for sound forest management among the decentralized and devolved systems. The DENR will have to agree on standards with the local government units, resource managers, community organizations, private sector and civil society. Meanwhile, the different stakeholders can start with previous and ongoing work on criteria and indicators of sustainable forest management and determine from this set which are worth tracking over time.

Box 11.3 Constraints raising the cost of doing business in the forestry sector

- weak tenure rights;
- unstable resource-use rights;
- inadequate budget support for local governments and DENR field operations;
- inadequate extension support for communities;
- no safety nets for displaced groups when their use rights are restricted;
- high cost of money and lack of long-term financing for forest plantations;
- unstable political conditions in areas suitable for forest production;
- agencies' overlapping mandates and convoluted regulatory systems, creating relatively high coordination and transaction costs;
- recentralized decision-making in issuing tenure and use rights; and
- poor infrastructure support in forested areas.

The local governments' major incentives are:

- establishing loyalty and support from the electorate in decentralized or devolved forests;
- introducing development with the objective of broadening local sources of revenues;
- getting people's support by facilitating resolution of conflicts; and
- participating in forest management decisions that may affect their use of forest resources within their jurisdiction.

Economic Values of Forestry

If managed well, forests and forest lands can yield not only private goods (such as profits) but very substantial public goods, a fact that is often overlooked when returns on investment are calculated. Biodiversity and other public benefits of well-managed forestry have economic value and should be considered in the equation.

Public goods generated by well-managed forests and forest lands, whether decentralized or devolved under different tenure systems, are expected to account for 20 to 80 per cent of the total estimated benefits (Table 11.1). Benefits from direct use values – private goods – range from 45 to 80 per cent (Francisco, 2004). Thus, if communities, indigenous peoples, non-DENR agencies and the private sector protect and manage their forests, both public and private interests can be simultaneously served. On top of this, profits from the private sector are taxed, generating revenue for the government. This implies that policies and their

Table 11.1 *Estimated economic value per hectare, by ownership*

Tenurial arrangement	*Area (ha)*	*Use value*	*Indirect value*			*Non-use value*	*Total economic value*
			Watershed protection	*Biodiversity values*	*Carbon sequestration*		
CBFM	21,084	58,556 65%	5932 7%	5544 6%	17,076 19%	2295 3%	89,402 100%
IFMA	31,078	59,229 69%	3499 4%	6977 8%	15,688 18%	0 0%	85,392 100%
CADC	36,660	65,273 68%	2795 3%	7453 8%	14,584 15%	5343 6%	95,448 100%
Co-Mgt	26,936	86,934 83%	4638 4%	3045 3%	8940 9%	811 1%	104,368 100%
Protected Area	26,558	29,715 46%	5005 8%	23,389 36%	4299 7%	2434 4%	64,842 100%

Notes: Values shown are in Philippine pesos. CBFM = community-based forest management; IFMA = integrated forest management agreement; CADC = certificate of ancestral domain claim; Co-Mgt = co-management.
Source: Francisco (2004)

implementing rules and regulations should therefore minimize entry costs for local government units, communities and the private sector to invest in forests and practise sustainable forestry with social equity, safety nets, transparency, the rule of law and accountability.

Moreover, incentives should consider the dynamic interplay of capacity to meet forest management standards, positive externalities and entry costs. Nobody wins when policies weaken the foundations of decentralized and devolved forest management, as is the case when existing tenure instruments are cancelled, issuance of devolved tenure instruments is suspended or entry costs are raised. Recentralizing forest management takes away the incentives of local stakeholders and resource managers to invest and benefit both themselves and the public.

Most policy instruments for decentralization and devolution do not allocate sustained funding support or mechanisms to carry out the new forest management responsibilities. Only one of the eight protected areas receives specific funding; legislation creating the others merely reads, 'subject to availability of funds or approved General Appropriations Acts'. A realignment of DENR budgetary resources to support top priorities is required. Co-financing, co-investment or payments for environmental services are a long-term approach to the sustained support of devolved or decentralized forest management. Inadequate funding to conserve biodiversity or sustain the environmental services of watersheds reflects government's shallow commitment to protecting these areas to serve public goods – a case where the 'talk' is not consistent with the 'walk'.

Some local government units seem reluctant to accept the challenge of co-financing. The local government units suspect that the DENR simply wants them to assume the costs of devolved forest management. Thus what they will get in return for their investment should be clearly defined and communicated. Until the local government units realize and appreciate the value of managing their forest lands as their own assets, political will may be lacking to support communities, co-managed areas, protected areas and, to a certain extent, larger watersheds.

Another hurdle is the apparent animosity and mistrust between the DENR and local government units. The DENR appears afraid of losing its regulatory powers, and its failure to consult about land allocation decisions leads local government units to suspect that the agency is simply playing games.

Emerging Results and Trends

The nature and extent of decentralized and devolved forest management in the Philippines in the past 15 years has been shaped by incentives for local forest management, donor funds, advocacy efforts, poverty and increased upland migration, and insufficient government efforts to reforest degraded areas and protect the remaining natural forests. This section considers the emerging results and trends influenced by policies on decentralization and devolution.

Forest land allocation

The distribution and extent of allocated forest lands (Guiang and Castillo, 2005) have significantly changed over the past 20 years (Table 11.2). Before the EDSA event (the People Power movement), most forest lands were under the concessions of timber licence agreements – as much as 11 million hectares in the early 1970s. Today, timber licence and industrial forest management agreements cover a little more than a million hectares of forest lands. The area and coverage of protected areas, watershed reservations, community forests and ancestral domains has increased dramatically since 1986, as shown in Figure 11.1.

The responsibility and cost centres in the protection and management of forest lands have shifted to the government, communities and indigenous peoples. The private sector's coverage has been significantly reduced because of social justice and equity concerns and perceptions of abuse and exploitation. Because of the Philippines' current tight fiscal situation, however, combined with increasing poverty and increasing domestic demand for wood and wood products, most of these forest lands may not be under effective and sound forest management. Illegal logging (mostly small-scale seasonal timber poaching and timber harvesting for

Table 11.2 *Area and extent of forest lands, by allocation*

State allocation	*Percentage of total forest lands and unclassified areas*
Public goods (set-asides for protected areas and watershed reservations)	29% (>4 million ha)
Civil and military reserves	Minimal
Local government units	Very minimal
Communities (community-based forest management and ancestral domain areas)	36% (>5 million ha)
Private sector (timber licence agreements, industrial forest management agreements)	7% (>1 million ha)
Unallocated forest lands (no tenure)	21% (>3 million ha)
Not yet allocated	7% (>1 million ha)

Sources: FMB (2002); Guiang and Castillo (2005)

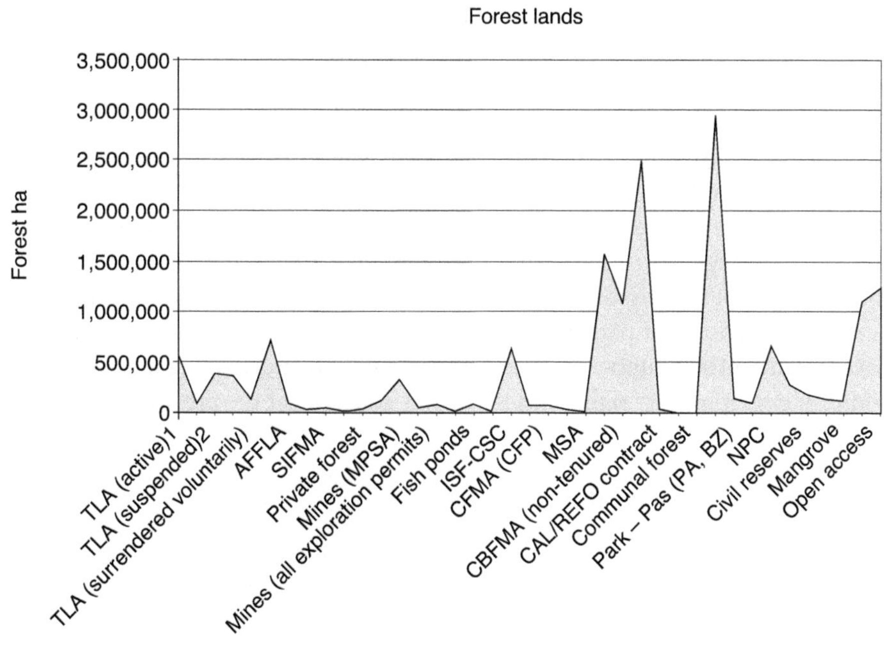

Figure 11.1 *A graphical view of forest lands allocation by tenure type*

Source: World Bank (2005b)

house construction and fuelwood) remains attractive compared with the cost (payoffs, delays, penalties, imprisonment) and difficult to control. The Philippines is also becoming increasingly dependent on imports and substitutes to meet its annual demand of about 5 million cubic metres of wood (Guiang, 2001). The private sector has been relying heavily on planted trees in private lands for their raw wood. According to Leonardo Angeles of the Philippine Wood Products Association, at least 60 per cent of the plantation timber that wood processors are now processing comes from private lands.

Based on an assessment of the Philippines Environmental Governance Project Phase 2 (EcoGov 2) of 210 tenure holders in 33 local government units, it was observed that, because of inadequate incentives, current fiscal deficits, declining subsidies and unpredictable forest policies, most community forests and ancestral domain areas, as well as DENR-managed protected areas and watershed reservations, are falling short of the minimum set of sound forest management indicators (Castillo and Guiang, 2006). The indicators include a resource management plan submitted to the DENR, a functional organization, an approved budget sufficient to cover the fixed costs of forest management (organizational, coordination and forest protection activities), a mechanism to address and formalize legitimate claimants within the area of the tenure holders (called 'individual property rights' and in the form of sub-agreements, certificates of stewardship contracts, fair sharing agreements for proceeds of timber sales, tribal council resolutions, and so on), mechanisms for resolving conflicts, livelihood programmes for communities, a working linkage with an external resource institution, and a programme to address gender concerns (DAI, 2005).

The current condition of decentralized and devolved forest management requires urgent action on the part of the DENR, tenure holders, local governments, donor agencies and civil society organizations. Immediate challenges are strengthening the forest management monitoring system with participation and enforcement mechanisms; providing or restoring incentives to community tenure holders; launching a national initiative with financial incentives that will induce more local government units to support forest management; and encouraging investment by addressing the property rights and incentives concerns of the private sector.

Poverty and reduced deforestation

Reductions in deforestation in the Philippines coincide with an increase in poverty. Despite the pump-priming in the forestry sector from development assistance funds (of almost US$1 billion), the upland poor are in a worsening situation (Figure 11.2). The designation of protected areas and watershed reservations has displaced upland people or restricted their use of forest resources for livelihoods. The government has not provided safety nets before setting aside large forest areas for biodiversity conservation and watershed reservations (Sachs, 2005). The resource-use rights to community forests have been suspended or cancelled at least four times since 1999, despite assessments showing that the few instances of exploitation or over-cutting occurred as a result of collusion with local forestry officials, inadequate implementation support to communities and over-regulation. The unpredictable and inconsistent government policies affecting the stability of property and use rights have discouraged the private sector from entering into joint ventures or subcontracts with forest-based communities.

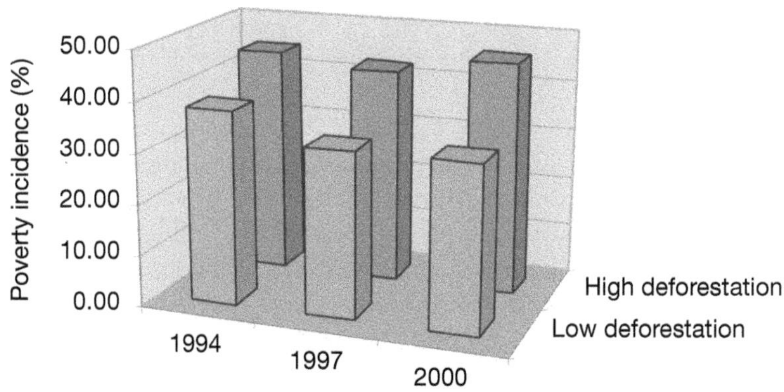

Figure 11.2 *Deforestation and poverty incidence in upland areas*

Source: NCSO (2002)

Gradual shift in land use and forest cover

The heavily regulated forestry sector has gradually shifted from harvesting natural forests to plantations of high-value trees and annual crops (DAI, 2004). Quoting a DENR estimate, the World Bank reported that, in 2002, the country's total forest cover increased by 24 per cent (World Bank, 2004) from a baseline of 5.4 million hectares in 1988. Recent updates of forest cover in Mindanao partly confirm this information (DAI, 2004) and show that, although the total natural forest cover decreased by an annual average of 41,000 hectares over the last 16 years, the area covered by tree plantations and perennial crops increased almost fivefold over the same period (Table 11.3). This is more than 78,000 hectares per year of annual increase. Demand for agricultural lands, population growth and to a certain extent the perceived lower economic value of natural forests accelerated the conversion of most open-canopy forest, brush lands and closed-canopy forests into plantations of fast-growing hardwoods and high-value perennial crops, including bananas, coconut, oil palm, rubber and fruit trees such as durian, mango and rambutan.

Table 11.3 *Forest cover in Mindanao in 1988 and 2004*

EcoGov category	*1988 SSC (ha)*	*2004 MMSE (ha)*	*Difference (ha)*
Closed-canopy forest	985,400	578,623	(406,777)
Open-canopy forest	1,823,900	1,572,946	(250,954)
Mangrove vegetation	46,200	116,375	70,175
Brushland and grassland in cultivated area	3,213,700	2,748,339	(465,361)
Tree plantations and perennial crops	271,000	1,524,111	1,253,111
Total	6,340,200	6,540,034	199,834

Source: DAI (2004)

Many private landowners are cutting their timber tree plantations and converting them into high-value tree crops because of the rising direct and indirect costs of harvesting and transport. Most private sector actors still hesitate to invest in forest plantations because incentives are not adequate (Acosta, 2004) and policies are often not consistent. Most plantings over the past 10 to 15 years have been financed with external loans and grants. Current regulations for harvesting planted trees by holders of stewardship contracts or community-based forest management agreements are discouraging upland farmers from planting trees; they are instead choosing to plant fruit trees and other high-value crops. This trend will mean that more forest lands are converted. Without adequate incentives to protect the remaining forests, except the fear of punishment or enforcement, forests whose management has been devolved to communities will either be illegally harvested for local household needs or will be harvested to supply makers of low-value furniture and similar products.

Partnership and Collaboration

Over the years, 'partnership and collaboration' has been the catchphrase of stakeholders in the Philippines' 'environment' industry (Morfit, 1998). The term is loosely defined as a 'working arrangement between those who have the mandate and those who are willing to support forest management and communities as de facto resource managers'. Partnership and collaboration occur as a result of informal or formal agreements between parties. The approach has been used to put together seemingly conflicting parties to confront common issues in environmental management. It has also been used to open up opportunities for other stakeholders, such as NGOs, religious groups, local leaders, indigenous peoples and the private sector, in the management boards, councils or steering committees of watershed reservations, co-managed areas, protected areas and water quality management areas.

Depending on the personalities and influence of those in these oversight bodies, these 'formalized coalitions' bring new insights, ideas and approaches to devolved forest management. They have become avenues to access external funding, advocacy, expert advice and political leverage. The downside, however, is the uncertainty of long-term funding. Protected area management boards, for instance, receive most of their support from local government units, grants, users' fees and the DENR. The coordination and meeting costs are enormous, especially with boards comprising more than 100 people.

Another approach involves provincial, regional and national federations of community-based forest management partners and holders of ancestral domain titles. Most initiatives in this area rely on external funding and will not be a major force unless local government units provide support and group members contribute from their own internally generated funds. The current free-rider attitude among such federations is not sustainable.

Some instruments of partnership and collaboration are listed in Box 11.4, with Box 11.5 showing the obligations in a typical memorandum of agreement between the local government units and the DENR in the joint implementation of approved forest land-use plans of municipalities or cities.

The partnership and collaboration between the DENR and the local government units have helped smooth many of the rough edges in joint protection, development and management of forest lands. The collaborative relationship could also open up opportunities for broadening the local government units' revenue base through the following:

Box 11.4 Philippine instruments of partnership and collaboration

- *memoranda of agreement* at the national level, normally mandated to organize committees for coordination and mobilization;
- *joint implementation memoranda of agreement* between local government units and the DENR for joint planning and implementation activities in forest lands or between two agencies for closure of open access, tenure assessment, provision of social infrastructure, conflict resolution, assistance to community-based forest management associations and ancestral domain title holders;
- *co-management agreements* that transfer direct management responsibility of specific forest lands to another party, such as a local government unit;
- *sub-agreements, outgrowers' contracts, business contracts and suppliers' contracts* between the board and committee chair, the legitimate forest occupants and the interested investors or supplier of services; and
- *resolution by the local government units for funding support* in the implementation and monitoring of joint forest management activities.

Box 11.5 Obligations in a typical memorandum of agreement between local government units and DENR

Joint obligations

- to assess tenure holders within the political jurisdiction of the local government unit;
- to close open access;
- to resolve boundary conflicts among different tenure holders and occupants;
- to promote public and private investments;
- to enforce forestry laws and regulations;
- to review applications for environmental compliance certificates; and
- to reclassify what have been classified as forest lands or allocate them to a different use, such as protected area, communal watershed or park.

Local government obligations

- to issue sub-agreements or recognize individual property rights within the tenured area;
- to facilitate public and private investments;
- to provide support systems, such as extension and infrastructure;
- to guarantee business contracts and financing;
- to pass resolutions to improve management of areas under sub-agreements or subcontract areas; and
- to impose fees.

- taxing occupied forest lands based on 'improvements over the area', as opposed to based on the land,
- encouraging private investment that will eventually generate tax revenue and ignite local economic activities;
- providing infrastructure in forest lands to stimulate markets, products, transport and productivity;
- institutionalizing payment for environmental services from heavy users of environmental services from forest lands either for water, ecotourism, bio-prospecting or high-value crop production, for example use of water in banana plantations, mariculture, and ponds or resorts; and
- reducing costs by hazard mapping and enforcing zoning based on watershed vulnerabilities to flooding and earthquake damage.

The Changing Roles of the DENR

In 1987, the forestry sector in the Philippines underwent a major change – institutionally and structurally (De Rueda, 2006). The Forest Management Bureau became a staff bureau and most of its former functions were transferred to the DENR regional offices. This shift signalled the increasing importance of national-level assessment and analysis as a basis for making sound policies and determining and defining standards and processes for sound forest management. This shift, however, has been slow in coming – a major reason for the current restructuring of the DENR bureaucracy.

Over the years, there have been a growing disconnect and overlaps in the functions of the Forest Management Bureau, Protected Area Management Bureau, Environmental Research and Development Bureau, and National Mapping and Resource Inventory Agency. All of these institutions are involved in forest management. Government restructuring is expected to address inadequacies and overlaps of these institutions and make them more effective, efficient and client-oriented. Draft recommendations for DENR restructuring have proposed significant reductions in field operations as more functions are devolved to the local government units (Malayang, 2005).

The forestry sector continues to experience increasing pressures and demands from its growing list of clients. The sector is not designed or structured to deliver service at the regional, provincial and community levels. New guidelines are expected to gradually transform the sector into a client-focused organization more willing to manage the forest based on multi-stakeholder perspectives.

Promising Practices in Forest Management

Current or recently completed foreign-assisted projects designed to support decentralized and devolved forest management have drawn on lessons from activities after the passage of the 1991 Local Government Code. These experiences demonstrate that, given the chance, partnerships between the DENR and local government units can work and that local governments – once they understand the benefits of taking charge of their resources – are willing to invest in forest management. Some models are emerging, and instruments have been developed to facilitate partnerships and thus help accelerate the devolution and decentralization processes.

Governance-oriented forest land-use planning and co-management

Forest land-use planning was originally a technically oriented process developed under the DENR's Natural Resources Management Program, an initiative supported by the US Agency for International Development (USAID). The process has been reconfigured as a tool for helping local government units, local stakeholders and the DENR agree on how forest lands should be allocated, managed with tenure holders, developed and protected. Plans are prepared after a thorough joint process of data gathering, analysis, validation and consultations between and among various stakeholders. In this process, the local government units are encouraged to think of forest lands not as 'cost centres' but as assets that can improve over time, sustain economic growth and serve the interests of local constituents. This improved version of the forest land-use planning with local government units is proving useful in bringing order to the allocation of forest lands. Now widely promoted and utilized by USAID's Philippine Environmental Governance (EcoGov) project in its implementation activities, forest land-use planning has produced a number of positive results.

Once the executive and legislative branches of the local government have reviewed and endorsed the plan, the DENR regional office approves it for implementation. The implementation agreement specifies the functions, obligations and roles of both parties. The local government can use the plan as a blueprint to plan and invest in farm-to-market roads, facilitate issuance of individual property rights by each tenure holder, provide a buffer for stabilizing property rights, provide needed support systems (extension, inputs, training, social infrastructures, linkage to financing and markets) to community tenure holders, and promote or invest in ecotourism facilities. The local government thus becomes the de facto provider of extension services, which are inadequately delivered by the DENR under its present structure.

During its first phase of implementation, the EcoGov project facilitated a governance-oriented and demand-driven forest land-use planning process with 33 local government units, and 21 completed the process. As of 2003, those 33 local governments had contributed about 15 million pesos (roughly US$300,000) from their own budgets to the planning process. Their financial contributions indicate their willingness to participate in the allocation, protection, development and management of the forest lands within their jurisdictions (DAI, 2004).

To date, 11 local government units in Central Visayas have planned to invest at least 128 million pesos (about US$2.5 million) to increase the asset value of their forest lands, totalling around 65,000 hectares. This is an average of 2000 pesos (US$40) per hectare. The funds will finance planting materials, capacity-building for communities, road improvements, extension services and forest protection. The DENR's planned budgetary support for these local governments' forest management is minimal.

The province of Nueva Vizcaya demonstrated how co-management of specific forest lands could be implemented. The local government unit issues sub-agreements to claimants and occupants inside its 24,000ha co-managed area to provide security of land tenure. Local ordinances and DENR support were established to implement the co-management agreement. The province and two municipalities have shown their support by providing planting materials (including fruit and other trees) to upland farmers and issuing agroforestry land management agreements, which have resulted in significant reduction of forest fires, charcoal making and timber poaching, increases in natural regeneration, and reduction of poverty in the province from 52 per cent in 1992 to almost 11 per cent in 2000 (see Agbayani, 2005).

Compared with the DENR's budgetary resources, local government units have tremendous potential to support decentralized and devolved forest management. In 2005, the total DENR budget (excluding the Environmental Management Bureau, Mines and Geosciences, and attached agencies) was only 4.556 billion pesos (roughly US$90 million), 80–85 per cent of which goes on operating expenses (World Bank, 2003). The bulk of the budget is allocated to the DENR's upkeep – salaries, rents and utilities. This means that less than a billion pesos (about US$20 million) was made available for the operations of the forestry, protected area systems, and research and development sectors (GAA, 2006). In contrast, the local government units (cities, municipalities and provinces) in 2005 had a total internal revenue allotment of 114 billion pesos (about US$2.28 million). Of the 114 billion pesos, 22 billion (roughly 20 per cent) may be used by the local government units for development, including the provision of environmental services. Funding to support environmental management services, especially for devolved forest management, could easily triple or quadruple financial resources at the local level. It should be noted that at least 40–80 per cent of local government units' total area is considered public forest and forest land, and more than 80 per cent of the country's land area is part of a watershed.

Subsidies for devolved forest management

The Community-Based Resource Management Project, supported by the World Bank, has made available more than US$50 million to subsidize investment in natural resources management by local governments and their communities. The rationale involves the interplay of poverty and resources: sound environmental management of forests and coastal resources would benefit not only the users but also the public in general. Funds are being managed by the Department of Finance, and the DENR's major responsibility is to assist local government units to identify and implement priority projects. Poor local government units can obtain a combination of loans and grants up to 70 to 80 per cent of the total amount of their proposed projects. To date, the experiences and lessons from this project have been positive. Local governments see the importance of investing in their forests and coastal resources as natural resource assets (World Bank, 2005a).

The Community-Based Resource Management Project has demonstrated the importance of subsidizing improvement of forests and coastal areas to arrest degradation and sustain these resources. Expanding this approach from its present regional coverage may be difficult, however, because of the current fiscal deficit. Early implementation also suggests that systems and processes should be further simplified and focused on forests and coastal resources that are of high priority for production, livelihoods, and protection of biodiversity and watersheds.

Policies that devolve forest management only partially have been another challenge in implementation. This has made the local government units the DENR's partners and clients in addressing decentralized and devolved forest management in the Philippines, and there is no turning back. Like the EcoGov project, the Community-Based Resource Management Project has shown that to manage forests and coastal areas, local governments and communities need long-term financing and sufficient technical capacities.

Examples from Ongoing Initiatives

In the province of Sarangani in Central Mindanao, the municipality of Maasim (covering 46,600ha of forest lands) is now reaping the fruits of its forest land-use plan, which was prepared with assistance from EcoGov and the DENR and is now being implemented with close collaboration between the local government unit and the DENR. Developing and completing the plan boosted the local government's confidence in being able to manage its own resources, which in turn boosted the confidence of investors. Dole Philippines has invested about 100 million pesos, engaging upland farmers in contract growing of pineapples in a 200ha area and leasing another 500ha for its own plantations. The contract growing arrangement is benefiting 200 households, a number of which had previously been communist rebels or supporters. More farmers are being employed as workers for both the contract growers and the company-run plantations. The DENR is helping the municipality monitor the plantation activities to ensure that these will not harm the ecological balance in the area.

The municipal government of Maasim allocated more than 500,000 pesos (about US$10,000) for forest land-use planning implementation in 2006 alone. It also obtained funding from various sources amounting to about 22 million pesos (roughly US$0.44 million) to build a 50km road network to improve access of upland dwellers to services and markets in the lowlands. What used to be unproductive land is now being turned into pineapple plantations, providing local communities with livelihoods.

Convinced that developing forest plans can stimulate economic growth, the Sarangani governor has signed memoranda of agreement with all its municipalities pledging commitment and financial support to forest land-use planning. For 2006, the provincial government has set aside 615,000 pesos (roughly US$12,300) for planning activities. This is in addition to its support for community forestry people's organizations, which will soon get subsidies in the form of seedlings valued at more than 200,000 pesos (about US$4000).

In all these activities, the DENR is an active partner, providing guidance and technical support. This partnership is proving beneficial not only to the local government units but to the DENR as well, since the agency can now perform its tasks with financial and manpower assistance from the local governments.

The Maasim and Sarangani cases are not isolated examples. In Central Visayas, particularly in Negros Oriental's Bayawan City, another EcoGov site, the local government is working closely with the DENR and various people's organizations to manage its watersheds and protect its sources of water. After consultations with the communities, preparation of maps and workshops, the forest land-use plan was completed and approved in December 2003. The following year, the city and the DENR signed an agreement to jointly implement the plan. The local government unit allocated 10 million pesos (about US$0.2) in 2006 for related activities.

With the money, the city acquired a brand-new surveying instrument for its forest lands. Both DENR and the local government team used the instrument in facilitating the identification and delineation of individual claims in the public forest lands and in ancestral domain areas. Other joint activities include a deputation of 92 ENROs (Environment and Natural Resource Officers), the majority of whom are officials of rural *barangays*; conduct of tenure assessments; and training of holders of community-based forest management agreements in resource management.

The money is also being used to rehabilitate portions of its watersheds. At the time of writing, the city has almost completed a 170,000m riverbank stabilization and rehabilitation project on the Bayawan with the help of several people's organizations. The city has planted 16,000 tree seedlings in a 24ha area, more than 800 fruit trees in a 5ha agroforestry area and 300 trees along a road. Another 10,000 seedlings were to be planted in 2006, and new nurseries will soon be established to ensure a ready supply of seedlings for the future.

Both Maasim and Bayawan have entered into co-management schemes with the DENR, with 12,000ha and 14,000ha of forest land, respectively. These local initiatives demonstrate the growing willingness of local governments to assume control of their forests as part of their natural resource assets.

MOVING FORWARD

There will be a continuing increase in the domestic demand for wood for construction, fuelwood, furniture and high-value products. As documented elsewhere (De los Angeles and Oliva, 1996; Guiang, 2001; FMB, 2005), domestic demand has in the past been met by imports, substitutes, plantations and illegal sources. Local governments and communities with large tracts of forest lands, working in partnership with local private sector actors and financiers, could easily seize this market, especially in Mindanao, Central Visayas, and parts of Southern and Northern Luzon, where agroclimatic conditions are ideal for forest plantations.

The local economic impacts of supplying wood will be enormous. The DENR should consider the market demand and bring down the cost of entry as much as possible. To this end, it should seriously consider giving community-based forest management areas and ancestral domain title holders monopolistic rights to harvest, process and market timber and non-timber forest products from productive secondary natural forests, with harvesting rights and volume based on their fixed costs and needs. This way, proceeds from the harvesting rights will include the cost of forest protection and rehabilitation, coordination and management of the people's organization.

Incentives and support for communal tenure holders, holders of individual property rights in the communal tenured areas, and interested private groups willing to enter into joint ventures or outgrowers' arrangements are the key to promoting investment. These stakeholders, given local government support and deregulation by the DENR, hold the key to future supplies of wood and fibre in the Philippines.

As municipal governments take centre stage in devolution and participate as partners in decentralized forest management activities, they should advocate for equity and a corresponding increase in their internal revenue allocations. While cities are getting an average of 1000 pesos per capita per year, municipalities are allocated only an average of 500. Municipalities involved in decentralized and devolved forest management are protecting and managing set-asides, such as protected areas and watersheds that serve the public good. They deserve a higher share of the national wealth to compensate for their efforts to protect and manage biodiversity and watersheds that greatly benefit downstream users, providing biodiversity, water, energy and aesthetic values. Local government units should also explore opportunities to obtain payments for environmental services.

There are opportunities for the government to broker long-term financing for tree crops, especially for slow-growing species, but this can be effective only if tenure holders have secure property rights, less regulation and lower entry costs for this kind of investment, and if the local governments, along with the national government, build infrastructure to support marginalized communities in forest lands. It has been observed, for instance, that in highly productive and cultivated forest lands, improved farm-to-market roads alone could reduce transport costs of products and inputs by about 50 per cent (World Bank, 2001).

Community forest and ancestral domain tenure holders have long wanted partners to help them plan and implement 'social enterprises' – enterprises that have the combined objectives of generating revenues, ensuring sound forest management, and achieving social justice and equity. The Philippines' 24 million residents in the uplands would surely welcome this kind of support from non-governmental organizations and firms with corporate social responsibility commitments. In most cases, however, these marginalized groups are treated as objects of research or as targets for justifying development assistance grants and loans from donors.

The threats to the Philippines' biodiversity remain a complex issue. The government and the international community have stakes in conserving biodiversity at the genetic, species and landscape levels for the benefit of present and future generations. The decentralized protection and management of protected areas and watersheds demand sufficient funding support, incentives for managers, incentives for communities, and long-term commitments from civil society groups and local governments. To date, the funds distributed to more than 400 protected areas have been considered insufficient to meet the minimum set of requirements for biodiversity reserves. Thus, releasing these areas for co-production, co-management, forest plantations and other uses may partly solve funding woes.

Lastly, the increasing shift toward decentralized and devolved forest management requires standards, systems and processes that are adaptable and usable at the local levels. These include the need for local governments to work together with the DENR in setting up a monitoring and evaluation system to assess periodic performance of forest tenure holders. Such a system, based on a few relevant indicators, should be used not just for compliance but to encourage better forest management.

References

Acosta, R. T. (2004) 'Impact of incentives on the development of forest plantation resources in the Philippines', in T. Enters and P. B. Durst (eds) *What Does it Take? The Role of Incentives in Forest Plantation Development in Asia and the Pacific*, RAP Publication 2004/27, Food and Agriculture Organization of the United Nations, Regional Office for Asia and the Pacific, Bangkok

Agbayani, R. (2005) 'Best practices in forest management through the co-management approach: The case of Nueva Vizcaya', paper presented at Philippine Forestry Development Forum, 2–3 June, Asian Development Bank, Manila

Bacalla, D. (2006) 'Community-based forest management: Policy changes and development', paper presented at Ten-Year Review of the CBFM in the Philippines: Forum for Reflection and Dialogue, 20–22 April, International Institute for Rural Reconstruction, Silang, Cavite, Philippines

Brillantes, A. (2000) 'Decentralization and devolution in the Philippines', in *Rural Development and Natural Resource Management: Trends, Strategy Implementation and Framework Performance Indicator System*, joint report of the Government of the Philippines and the World Bank, Manila

Borlagdan, S., Guiang, E. S. and Pulhin, J. (2001) *Community-based Forest Management in the Philippines: A Preliminary Assessment*, Institute of Philippine Culture, Ateneo de Manila University, Quezon City, Philippines

Castillo, G. and Guiang, E. S. (2006) 'Report of the tenure assessment in EcoGov2 – Assisted local government units', DAI/EcoGov 2 Office, Pasig City, Philippines

Contreras, A. (ed) (2003) *Creating Space for Local Forest Management in the Philippines*, La Salle Institute of Governance, La Salle University, Manila

Coronel, S., Chua, Y. T., Rimban, L. and Cruz, R. (2004) *The Rulemakers: How the Wealthy and Well-born Dominate Congress*, Philippine Center for Investigative Journalism, Quezon City, Philippines

Dacanay, M. L. (ed) (2004) *Creating Space in the Market: Social Enterprise Stories in Asia*, Asian Institute of Management and Conference of Asian Foundations and Organizations, Makati City, Philippines

DAI (Development Alternatives Inc) (2002) 'Autonomy and devolution: Proceedings of EcoGovernance round table discussion', EcoGov 1 reports, USAID, Manila

DAI (2004) 'Updating the forest cover map in Mindanao using satellite imagery', DAI technical report, EcoGov 1 Project, USAID, Manila

DAI (2005) 'EcoGov 2 project performance monitoring plan', DAI/EcoGov 2 Project, Pasig City, Philippines

De los Angeles, M. (2000) 'Natural resource management: Trends, strategies and performance indicators', in *Rural Development and Natural Resource Management: Trends, Strategy Implementation and Framework Performance Indicator System*, joint report of the Government of the Philippines and the World Bank, Manila

De los Angeles, M. and Oliva, R. (1996) 'Logging bans: Economic and environmental consequences', Natural Resource Management Program/Development Alternatives Inc, DENR, Quezon City, Philippines

De Rueda, R. (2006) 'Re-inventing the forestry agencies of the Department of Environment and Natural Resources: Philippines', case study prepared for FAO, Bangkok

FMB (Forest Management Bureau) (2002) 'Forestry statistics', Forest Management Bureau, DENR, Quezon City, Philippines

FMB (2005) 'Forestry statistics', Forest Management Bureau, DENR, Quezon City, Philippines

FMB (2006) 'Draft omnibus forestry guidelines', Forest Management Bureau, DENR, Quezon City, Philippines

Francisco, H. (2004) 'Economic analysis of allocating forest and forest lands: Total economic valuation approach', technical report submitted to EcoGov 1, Quezon City, Philippines

GAA (2006) www.dbm.gov.ph

Guiang, E. S. (2001) 'Efficacy of removing natural forest from timber production as a strategy for conserving forests in the Philippines', in P. Durst, T. Waggener, T. Enters, and T. Cheng (eds) *Forests Out of Bounds: Impacts and Effectiveness of Logging Ban in Natural Forests in Asia-Pacific*, FAO, Bangkok

Guiang, E. S. and Castillo, G. (2005) 'Trends in forest ownership, forest resources tenure and institutional arrangements in the Philippines: Are they contributing to better forest management and poverty reduction?' case study prepared for FAO, Bangkok

Magno, F. (2003) 'Forest devolution and social capital: State–civil society relations in the Philippines', in A. Contreras (ed) *Creating Space for Local Forest Management in the Philippines*, La Salle Institute of Governance, La Salle University, Manila

Malayang, B. (2005) 'Recommendations for devolving ENR functions to the local government units', technical assistance report for the EcoGov 2 Project, League of Municipalities of the Philippines, and Department of Interior and Local Government, DAI EcoGov 2 Office, Pasig City, Philippines

Mercado, E. (2000) 'Decentralization and devolution of forest management in the Philippines: Uneasy steps to institutional maturity', in T. Enters, M. Victor and P. Durst (eds) *Decentralization and Devolution of Forest Management in Asia and the Pacific*, RECOFTC Report No 18 and RAP Publication 2000/1, Regional Community Forestry Training Center, Bangkok

Morfit, M. (1998) 'Framework for DENR–local government unit strategic partnership for natural resources management', Natural Resources Management Program/ Development Alternatives, Inc (NRMP/DAI) technical report, DENR, Quezon City, Philippines

NCSO (National Census and Statistics Office) (2002) *Philippine Statistical Yearbook*, NEDA, Makati City, Philippines

Porter, G. and Ganapin, D. (1988) 'Resources, population and the Philippines' future: A case study', Paper 4, World Resources Institute, Washington, DC

Pulhin, J. (2006) 'Where goest thou CBFM?', paper presented at Ten-Year Review of the CBFM in the Philippines: Forum for Reflection and Dialogue, 20–22 April, International Institute for Rural Reconstruction (IIRR), Silang, Cavite, Philippines

Sachs, J. D. (2005) *The End of Poverty: Economic Possibilities for Our Time*, Penguin, London

Vitug, V. (1993) *The Politics of Logging: Power from the Forest*, Center for Investigative Journalism, Manila

World Bank (2001) 'Aide memoire during the World Bank's supervision mission of the Agrarian Reform Communities Development Project', World Bank of Manila, Pasig City, Philippines

World Bank (2003) 'Governance of natural resources in the Philippines: Lessons from the past, directions for the future', Rural Development and Natural Resources Sector Unit, East Asia and Pacific Region, World Bank of Manila, Pasig City, Philippines

World Bank (2004) 'Philippines: Environment monitor 2004', World Bank of Manila, Pasig City, Philippines

World Bank (2005a) 'Aide memoire during the World Bank's supervision mission of the Community-Based Resource Management Project', World Bank of Manila, Pasig City, Philippines

World Bank (2005b) 'Natural resources management: Way forward action plan for the Philippines', Rural Development and Natural Resources Sector Unit, East Asia and Pacific Region, World Bank of Manila, Pasig City, Philippines

12

The Household Economy and Decentralization of Forest Management in Vietnam

Tan Quang Nguyen

In recent years, the government of Vietnam has taken the bold step of devolving forest management rights to local people. What are the implications for households? This chapter subjects the question to empirical study in Dak Lak Province, Vietnam, using field data collected in 2002–2003 in 13 villages where forest devolution took place between 1999 and 2000. Findings suggest that forest devolution has the potential to contribute to the household economy, but to varying degrees across households and villages. Consequently, the incentives for local people to participate in the management of devolved forest depend on having a feasible benefit-sharing mechanism that addresses the needs of the poor and disadvantaged members of the community.

Historical Background

The past two decades have seen radical changes in forest tenure arrangements in Vietnam. The changes were driven by a quick decline in the area of forest a decade after unification, the ineffectiveness of the state forest enterprise system in forest management, the state's increased recognition of the importance of local people in forest management and the success of agricultural reforms of the late 1980s. Now the government is devolving forest management rights. Furthermore, the new Forest Protection and Development Law, passed in December 2004, contains clauses in favour of community forest management. The ultimate purpose of such policies is not only to maintain and improve the existing cover of forest but also to improve forest people's participation and livelihoods. For this purpose, it is important to know how forest devolution policy is affecting the economy of local households before it is applied to the whole country.

Forests and Forest Management in Vietnam

Vietnam is a tropical country located in Southeast Asia's Indo-China Peninsula. Its territory stretches from 8°02' to 23°23' North latitude and from 102°08' to 109°28' East longitude. The total land size is 33.038 million hectares, administratively divided into 64 provinces and municipalities.

According to the most recent statistics, 57.6 per cent of the land in Vietnam is classified as forest land (Table 12.1). Total forested area is 12.62 million hectares, of which natural forest covers 10.28 million hectares and plantation forest 2.33 million hectares. This is equivalent to a national forest cover of 38.2 per cent (31.1 per cent natural forest and 7.1 per cent plantation forest). By uses, forests in Vietnam are classified as production forests (36.3 per cent of the total forested area), protection forests (48.1 per cent) and special-use forests (15.6 per cent).

Table 12.1 *Forest classification according to uses*

Forest types	*Total area (ha)*	*Special-use forest (ha)*	*Protection forest (ha)*	*Production forest (ha)*
I. Forested area	12,616,700	1,958,320	6,172,062	4,486,318
A. Natural forest	10,283,173	1,874,829	5,302,652	3,105,693
B. Plantation forest	2,333,526	83,492	869,410	1,380,625
II. Non-forested forest area	6,411,990	417,716	3,377,417	2,616,857
Total	*19,028,690*	*2,376,036*	*9,549,479*	*7,103,175*

Source: www.kiemlam.org.vn

At present, forest land in Vietnam is being managed under different tenure arrangements. According to the current definition applied by the Ministry of Agriculture and Rural Development, forest owners comprise eight major groups: 1) state enterprises, 2) management boards of protection forest (MB-PFs), 3) management boards of special-use forest (MB-SFs), 4) joint venture enterprises, 5) individual households, 6) collectives, 7) army units and 8) people's committees. Among these owner groups, state enterprises, individual households and people's committees are the largest forest owners, and army units and joint venture companies own the least forest area (Figure 12.1).

Forest management under the eight ownership groups can be classified into four major tenure arrangements: public property, private property, common property and forest contracting (see also Nguyen, 2005a). State or public property refers to forests under management by owner groups 1, 2, 3, 7 and 8 above. Under

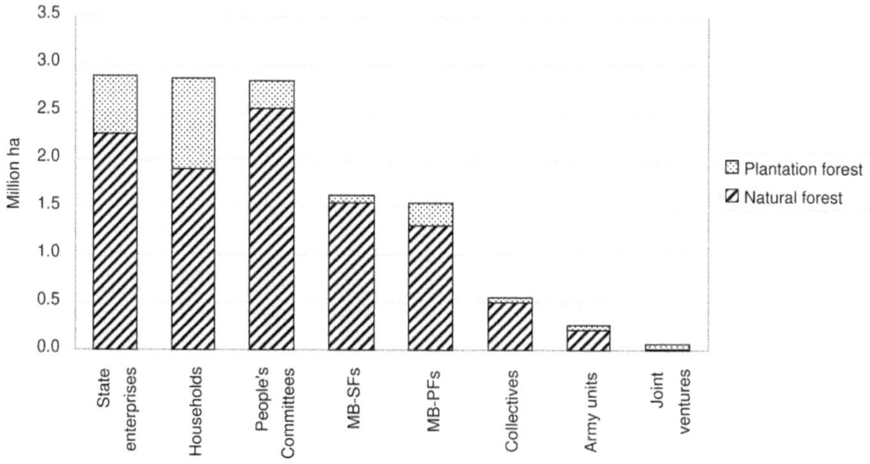

Figure 12.1 *Forest area by ownership group*

Note: Data sorted in descending order.

Source: www.kiemlam.org.vn

this arrangement, forest is allocated to state bodies for an unspecified period of time. Where the forest falls into the special use or protection category, forest owners are entitled to receive state funds for management of the forest. Private property includes forest management by individual households and joint venture enterprises (owner groups 4 and 5 above). Under this arrangement, forest is allocated to its owner for long-term (50-year) management. In most cases, forest owners under this arrangement are entitled to a legal title for the forest area they are given. Common property is found in forest managed by collectives (owner group 6). However, official statistics count only those collectives legally recognized by the state; areas under traditional communal management (not officially recognized by the state) are not included. Under the collectives system, forest is allocated to a group of individuals, each with similar rights and responsibilities. Finally, forest contracting in Vietnam is formed when an owner of state forest property signs a contract with an organization, a household, a group of households or a village to protect the forest. Under this arrangement, rights of ownership of the forest under contract remain with the contractor and the contractee has only the rights specified in the contract.

The area under different forms of ownership is shown in Table 12.2. The area of forest under contractual arrangement is included as part of the area under state ownership. At present, there are around 2.26 million hectares (17.93 per cent of the total forest area or 24.76 per cent of the forest under state ownership) under protection contracts with local people (see also www.kiemlan.org.vn; Nguyan, 2005a).

Table 12.2 *Forest area by tenure arrangements*

	Natural forest		*Plantation forest*		*Total*	
Ownership	*Area (ha)*	*Percentage*	*Area (ha)*	*Percentage*	*Area (ha)*	*Percentage*
Public	7,862,663	76.5	1,273,054	54.6	9,135,717	72.4
Private	1,919,472	18.7	1,002,041	42.9	2,921,513	23.2
Common	501,038	4.9	58,432	2.5	559,470	4.43
Contract					2,261,966†	

† This figure is from the Five Million Hectare Reforestation Program only; the area is formally state property.

Sources: www.kiemlam.org.vn; Nguyen (2005c)

Decentralization of Forest Management

Changes in forest policy are summarized in Box 12.1 (see also Nguyen et al, 2001; Nguyen, 2005a). Until the early 1990s, management of forest by state organizations was the only form of forest tenure. Efforts were made to introduce private forest management in Vietnam following the approval of the Forest Protection and Development Law in August 1991. In addition to introducing 'private property' in forests, the law also established a legal basis for setting up management boards for protection forests and special-use forests as new forms of state forest management. Private management of forest land was also supported with the approval of the new Land Law in July 1993, according to which land users were entitled to long-term (50 years for forestry land and 20 years for agricultural land), renewable land-use titles, or Red Book certificates, with five rights: to exchange, to transfer, to inherit, to mortgage and to lease.

In 1994 and 1995, people's involvement in forest management was supported by two decrees. One, intended to allocate forest land to organizations, households and individuals for long-term (50-year) use in accordance with the use of each forest type (production, protection or special-use forests), provided a framework for transferring management of forest land from the state to local organizations, households and individuals. The other supported the involvement of people in forest management through long-term contracts with state organizations.

However, only owners of non-forested land or plantation forest were given Red Book certificates; no natural forests were allocated. A subsequent decree, issued in November 1999, provided a new framework for allocation and lease of forest land to organizations, households and individuals for sustainable and long-term forestry uses. Though the purpose of allocation remained similar, it was now possible to issue titles for allocated forest land.

By the end of the 1990s and early 2000s, forest management by household groups or communities emerged. Though community forest management was not

Box 12.1 Major milestones in forest decentralization policies in Vietnam

December 1986: *Doi Moi* (economic renovation) policy launched after the determination of the Sixth National Congress of the Vietnamese Communist Party.
August 1991: Law on forest protection and development passed by the National Assembly, marking an effort to involve local people and different economic sectors in forest protection and development.
July 1993: Land law passed by the National Assembly, stipulating the rights of landowners to lease, exchange, inherit, mortgage and transfer land-use titles.
January 1994: Government Decree 02/CP on transfer of management of forest land from the state to local organizations, households and individuals.
January 1995: Government Decree 01/CP on the allocation of land through contracts for the purposes of agriculture, forestry and aquaculture.
November 1999: Government Decree 163/1999/ND-CP on land allocation for forestry purposes.
November 2003: Land Law passed by the National Assembly, recognizing the legal status of communities as owners of land resources.
October 2004: Government Decree 181/2004/ND-CP, providing a new framework for allocation of forest land (replacing Decree 163), following the ideas of the 2003 Land Law.
December 2004: Law on Forest Protection and Development passed by the National Assembly, recognizing common property as a legal forest management arrangement.
Mid-2006: Community forestry guidelines ready for trial.

a new form, this period marked an important change in the legal recognition of this management arrangement. The new Land Law, passed in November 2003, and the new Forest Protection and Development Law, approved in December 2004, provide legal recognition of 'ownership' of forest and forest land by communities. An October 2004 decree provided a new framework for allocation of forest land, to implement the 2003 law. In mid-2006, guidelines for implementation of community forest management were drafted and trials began in 10 provinces to assess outcomes before the document becomes legally binding for the whole country. Once that is the case, procedures for devolution of forest to communities and community forest management will be based on this document.

To date, the most important outcome of tenure reform has been the involvement of various stakeholders, particularly local people, in resource management (MARD, 1999; Nguyen et al, 2001). Although currently the state still administers a large area of forest, the trend shows an increase in forest area under long-term management by people, either individually or collectively. For example, the forest area under management by local people has increased from 2.62 million hectares in 2003 to 3.16 million hectares in 2004 and 3.41 million hectares in 2005, or from 29.8 per cent to 34.7 per cent to 37.4 per cent of the forest area under state management in the respective years. Furthermore, a legal framework for tenure reform and benefit sharing now exists.

Nevertheless, challenges remain for decentralization of forest management in Vietnam (Nguyen, 2003; Wode, 2003; Nguyen, 2005b; Roth, 2005). In general, forest tenure reform is likely to be successful if the allocated land is near residential areas (in most cases this is barren land and poor forest). A problem often arises with management of natural forests, which are often relatively far from residential areas and thus difficult to protect and manage. In addition, forest boundaries are often uncertain and the immediate benefits are not apparent to the local people. Most people still consider natural forests – even if they have been allocated – 'open access' resources that belong to everyone. Furthermore, many people like to convert natural forest into plantations, which offer more immediate benefits. As a consequence, a benefit-sharing mechanism that is clear and secure for forest holders has yet to be achieved.

Forest Devolution in Dak Lak Province

Dak Lak and its forests

Dak Lak Province is situated in the Central Highlands of Vietnam (Figure 12.2). In 2004, Dak Lak was split into two provinces, but when the research for this chapter was conducted, it was still one province, with 1.96 million hectares, more than half of which was classified as forest land. Agricultural land accounted for 26 per cent of the total land area. The population was about 2 million people, most of whom had come from other provinces during the previous two and a half decades. The Ede, Jarai, Mnong and other indigenous groups accounted for around 18 per cent of the province's population.

At the end of 2003, Dak Lak had 979,000ha of forest (959,000ha natural forest and 20,000ha plantation), equivalent to 50 per cent forest cover. Almost half of the forested area was production forest. The remainder was split between protection and special-use forests.

Since the end of the American War (known in the West as the Vietnam War), forests across Dak Lak had been managed by the state. At the field level, state forest enterprises were in charge of forestry activities. Forest cover in Dak Lak declined rapidly after the war. Around 242,000ha of natural forest were lost between 1982 and 1999. In addition, forest quality decreased. Areas with high-quality forest declined from 73,000ha in 1982 to 15,000ha in 1999, and poor forest increased from 278,000ha to 411,000ha in the same period (DARD, 2001b).

Figure 12.2 *Dak Lak, Vietnam*

Experimental forest devolution

In response to the deforestation and degradation of forest resources, Dak Lak initiated an experimental forest devolution programme. The initiative started in 1998 with technical support from a German-funded project, namely the Sustainable Management of Natural Resources in the Lower Mekong Region Project. The objectives of the programme were 1) to maintain and improve the province's forest cover and 2) to improve the livelihoods of local people. The idea was to devolve the rights to natural production forests from local state forest enterprises to either individual households or groups of households. Beginning in 2001, the programme also handed over forest management rights to communities (Nguyen, 2005b).

Dak Lak's forest devolution programme fitted into the province's political economy. More than two decades of state forest management, coupled with the recent development of cash crops and the arrival of migrants, had led to the erosion of the local land tenure system (Bui, 1989; Ngo, 1989; Nguyen, 1989). By devolving the rights to local forests to the local inhabitants, particularly to the indigenous groups, Dak Lak's forest devolution programme called attention to the indigenous land tenure system. Though the programme still emphasized state control over local forest resources, it marked the start of a change in the province's policy framework – towards recognition of local land tenure.

In its experimental programme, Dak Lak granted long-term land-use titles (Red Book certificates) for forested land to local people. In addition, a contractual agreement between the state forestry representative, local authorities and local people was attached to each title. This contract specified the benefits from devolved forest that people were entitled to. Nevertheless, the programme did not specify what mechanisms were available or how to realize these benefits. Furthermore, the contract set constraints on what people could and could not do with the devolved forests. For example, forest recipients were required to ask for permission from competent state organizations to collect timber products or use forest land for cultivation. The restrictions did not conform with the rights vested in the land-use title as specified in the Land Law. Lacking prior experience, Dak Lak proceeded slowly and gradually introduced the concept of individual 'ownership' of natural forest into the existing forest tenure arrangements.

At the field level, the programme started with a plan to devolve 7000ha of forest from state forest enterprises to local people in 1999. By the end of 2000, forest devolution (in the field) was completed in 13 villages, with approximately 7100ha of forest having been devolved to 339 individual households and to 19 household groups consisting of 149 households (DARD, 2001a). Issuance of forest land-use titles to local forest managers was completed shortly thereafter. In April 2001, the province reviewed the forest devolution process, after which the programme continued with an experiment in the third form of devolution: forest devolution to a community. By the end of 2002, the programme had devolved 24,900ha of forest land to 3243 individual households, 10 groups of 111 households, and 24 communities (including the 13 villages in which forest devolution had been completed in 2000). To date, an additional area of around 7700ha has been allocated to 18 household groups and seven communities. On average, an individual household received 5–10ha of forest, a group of households 150–200ha and a community 400–500ha (Nguyen, 2003).

In general, the area of forest devolved to local people was modest. Forest devolution in Dak Lak has, however, made two breakthroughs. First, it gave local households the rights to natural forest resources together with the Red Book certificates for forested land. Second, it reintroduced the ideas of direct sharing of timber benefits and land use for cultivation purpose by local forest managers. These two points had been under discussion before, but Dak Lak's forest devolution programme was the first to effect change. As a result, the programme has gained the attention of international donors, forestry policymakers from the ministry level and other provinces (Nguyen, 2005b; Tran, 2005). Learning from the Dak Lak example, Son La Province in the Northwest also took action to devolve its forest area to local people with the help of the German-funded Song Da Social Forestry

Development Project. By the end of 2004, around 825,000ha of forest in Son La had been devolved to local people.

It is still not clear, however, whether the forest devolution programme in Dak Lak has achieved its stated objectives. A 2003 assessment of 10 villages revealed that forest cover and volume of standing timber were still declining (Tran et al, 2003b). In addition, benefits from the forest tended to increase after devolution but their distribution among local households and villages was skewed. As discussed by Nguyen (2005b and 2006b) and Tran et al (2003b), factors hindering the fulfilment of intended objectives include inadequate attention to local rules on forest tenure, lack of legal support to help local people realize their new rights, unclear and unfeasible benefit-sharing mechanisms, and insufficient or no support to help local people, particularly the poor and disadvantaged, benefit from the forest.

Research Setting and Methodology

By the end of 2000, when the experimental forest devolution programme in 13 villages was almost complete, Dak Lak asked the Tropical Ecology Support Program of the German Agency for Technical Cooperation (GTZ) to support research to evaluate outcomes at the local level. The proposal was accepted, and the Dak Lak Department of Agriculture and Rural Development took charge of conducting the research with academic support from Humboldt University, Berlin, and institutional support from the Sustainable Management of Natural Resources in the Lower Mekong Region Project, and later the Watershed Management Component under the Mekong River Commission. Fieldwork was conducted from March 2002 to June 2003 in all 13 villages in five districts where forest devolution took place.

Of the thirteen study villages, four were dominated by Jarai people, five by Mnong and two by Ede. The two remaining villages consisted of migrants who had come to Dak Lak during the 1980s. Populations ranged between 290 and 590 people, and villages had 42 to 115 households. The villages were located in an economically poor area, according to Vietnamese classification. The local people's most important source of livelihood was agriculture. Forest products were collected mostly for domestic use. The most popular off-farm sources of income were salaries, war remuneration, and returns from trade and services.

The research sought to examine:

- harvests of major products from the devolved forest, which included timber, non-timber forest products (bamboo, bamboo shoots and forest leaves) and crop production on forest land;
- the relative economic significance of forest harvesting compared with other sources, such as cropping and off-farm income; and

- the contribution of forest devolution to access to credit by local households (see also Nguyen, 2005b, 2006a and 2006b).

Data were collected at the household level in two studies. The first study looked at two villages in two districts between March and September 2002. A census of all 95 households in the two villages was undertaken. In the second study, 228 out of the total 931 households in 11 villages were surveyed between February and May 2003. Households in the second study were selected based on a stratified random sample, in which all households in a village were stratified according to legal ownership (or not) of devolved forest and wealth, and then selected randomly. Altogether, data for 323 households from 13 villages were used (see Nguyen, 2005b, for a detailed discussion on data collection and methodology). Results are reported in Tran et al (2003a, 2003b and 2004), Sikor et al (2005), Tran (2005), Nguyen (2005b and 2006b), and Tran and Sikor (2006).

It is important to note that Dak Lak remains a special case of forest devolution in Vietnam because it received substantial technical and research assistance in implementation and documenting the process and its effects. Only one other province in the northwest of the country, Son La, has received a similar level of external technical and research support.

Economic Implications for Local Households

Potential for economic benefits

Under devolution, recipient households were entitled to Red Book certificates (which could be used as a mortgage against formal credit) and became 'owners' of the devolved forest. Households possessing certificates, whether jointly or individually, were legally entitled to specific uses of the forest resources; those who were not included in the devolution programme had no legal rights to the forest. Each recipient household was entitled to $10m^3$ of timber (originally half that) every 20 years for housing, with logging activity subject to approval by a competent state organization. In addition, when forest became mature enough for commercial logging, the household would be entitled to a 6 per cent share of the after-tax value of logged timber for every year of protection. Forest recipient households were entitled to collect and use all non-timber forest products under the forest canopy without having to pay the resource tax. Each household would also be legally entitled to a limited area of cultivated land in the devolved forest, the conversion of which was subject to approval. It was unclear, however, how much land a household would be legally allowed to use for agricultural purposes.

Of the surveyed households, 55 per cent (179 out of 323 households) were entitled to forest Red Book certificates and could legally enjoy the forest benefits described above. These households received a total of 2615ha of forest with an estimated standing timber volume of 153,827m^3. Average forest size per household with forest was 14.6ha, ranging from 3ha to 28ha. Average timber volume was 859m^3, ranging from 0 (no timber forest) in 23 households to more than 3500m^3. In theory, these timber and forest areas would be the potential benefits for the forest recipient households. Unfortunately, no data on non-timber forest products were available.

Harvests from devolved forests

Timber harvest

The survey asked local people to choose the three timber products that were most significant to them. All households in each village were then asked about the harvest of those products for the study years (1999 and 2001–2002). For the sake of comparison across sites, the results are presented in the products' cash value.

Timber appeared to be less important than other resources from the devolved forest (Table 12.3). In 2001, only 81 of the 323 surveyed households (25 per cent) had harvested timber products. Among them, 48 households had legal forest Red Book certificates and 38 did not. The value of timber harvest was between 0.08 million and 16 million Vietnamese dong (VND) per household, and about 2 million VND on average (US$1 ≈ VND15,000 in 2002–2003).

Table 12.3 *Harvests from devolved forests*

	Households (percentage of total)	*Total quantity*
Harvest of timber	81 (25)*	165.5 million VND
Harvest of non-timber forest products	145 (45)	18.6 million VND
Upland field in the devolved forest	119 (37)	90.56ha
Paddy field in the devolved forest	23 (7)	8.2ha
Crop harvest from field in devolved forest	110 (34)	161 million VND
Aggregated harvest from devolved forest	217 (67)	345 million VND

Note: * Number in parentheses refers to the percentage of households deriving the corresponding harvest over the total surveyed households.

Non-timber forest products

As for timber, people identified the three most important non-timber forest products, and the harvests were converted to a cash equivalence value. Non-timber forest products from devolved forests appeared to be more popular than timber, but their value was lower. Around 45 per cent of the surveyed households (145 out of 323, comprising 90 with legal certificates and 45 without) collected some 'considerable' quantity of these products from the devolved forest. The value ranged between 0.01 million and 2.7 million VND per household. An average household collected 0.13 million VND worth of non-timber forest products.

Cultivated land and crop production

The use of land in the devolved forest for cultivation was more important than timber but less important than non-timber forest products. Around 37 per cent of the surveyed households (119 out of the total 323, comprising 95 with legal certificates and 24 without) had a total of 90.56ha of upland field in the devolved forest. Fields ranged between 0.06 and 3.7ha, with an average of 0.76ha per household.

In addition to upland fields, 7 per cent (23 households – 18 with certificates and 5 without) had a total of 8.2ha of paddy field in the devolved forest. Fields ranged between 0.1 and 3ha, averaging 0.025ha per household. Around 13 of the 23 households who had paddy fields in the devolved forest also had an upland field.

Of the total surveyed households, 110 (86 with certificates and 24 without) had harvests from cultivated land (both upland and paddy fields) in the devolved forest. The value of the crop harvests was between 0.05 million and 9.8 million VND per household, averaging 1.4 million VND.

In summary, around 67 per cent (217 households) of the surveyed households had some kind of harvest from the devolved forest in 2001–2002, with a total cash equivalence value of 345 million VND (excluding cultivated land). Among them, 20 households harvested timber, non-timber forest products and crops; 79 households harvested two kinds of products; and 118 households harvested only one. Distribution of aggregated harvests ranged between 0.01 million and 16 million VND, an average of 1.6 million VND per household.

Harvests before and after devolution

As Table 12.4 indicates, total upland area in the devolved forest cultivated by surveyed households increased from 36.83ha in 1999 to 90.76ha in 2002, representing an increase of almost 250 per cent. By contrast, the change in paddy land was relatively minor. A statistical test confirms the significant difference in means of upland fields in the two years, with $t=4.0309$ and $p>t<0.0001$, and the insignificant difference in means of paddy fields, with $t=0.1549$ and $p>t=0.8769$.

A significant increase in harvests from fields in the devolved forest was also observed. Overall, there was a growth of 300 per cent in the value of crop outputs. Statistical tests confirm a significant difference in harvests before and after devolution, with $t=4.314$ and $p>t<0.001$.

By contrast, the difference in harvests of timber products from devolved forests before and after devolution was insignificant: the change in total output value was 120 per cent, and statistical tests show insignificant differences in the timber

Table 12.4 *Difference in benefits before and after devolution*

	1999	*2001–2002*	*Difference*	*Change (percentage)*
Upland field in forest (ha)	36.83	90.76	53.93	246%
Paddy field in forest (ha)	7.49	8.20	0.71	109%
Harvest from field (VND thousand)	52,566	161,359	108,793	307%
Timber (VND thousand)	136,983	165,480	28,497	121%
Nontimber forest products (VND thousand)	11,675.5	18,616	6,942	159%
Aggregated harvest (VND thousand)	201,225	345,456	144,232	172%

harvests of the two years, with t=0.7651 and p>t=0.4445. As for non-timber forest products, the difference in value of harvests before and after devolution was around 160 per cent. Statistical tests back up this fact, with t=1.7724 and p>t=0.768.

Nevertheless, the aggregated value of all harvests – timber, non-timber and crops – from the devolved forest was significantly larger following forest devolution than preceding it. An increase of over 170 per cent was observed between 1999 and 2001–2002 (Table 12.4). Results of statistical tests show a significant difference before and after devolution, with t=3.2239 and p>t=0.0013.

Economic significance of forest harvests

Forest harvests versus major sources of livelihoods

The harvests from devolved forests played an important role in the household economy (Figure 12.3). Compared with total harvests from cropping, the most important source of local livelihoods, the value of harvests from devolved forest was around 22 per cent (345 million VND of total harvest from devolved forest compared with 1.55 billion VND from crop production) in 2001–2002. In two

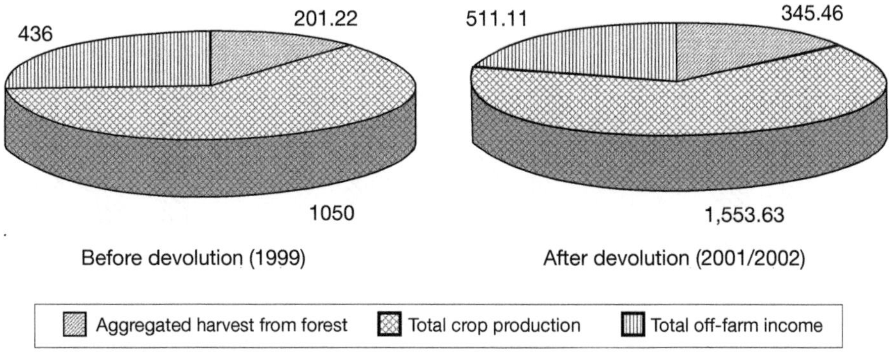

Figure 12.3 *Total harvest from forest versus off-farm income and crop outputs*

villages, harvest from devolved forest was more than 34 per cent of the total value of crop production; in a third village, it was 8 per cent; and in the remaining villages, it ranged between 14 and 26 per cent.

Compared with aggregated income from off-farm sources, the total value of harvests from the devolved forest accounted for an average of 67.6 per cent. In six villages, aggregated harvests surpassed cash income from all off-farm sources. By contrast, value of the harvests was 30 per cent or less in three villages. It was between 47 and 68 per cent in the remainder.

The contribution of harvests from devolved forests to the household economy grew not only in absolute numbers but also proportionately compared with income from off-farm sources and total crop production. From 1999 to 2001–2002, harvests from forest grew 22 per cent relative to total off-farm income and 3 per cent relative to crop production.

Paddy and upland fields in forests versus total farmed land

Cultivated land in the devolved forest represented an important portion of household land resources (Figure 12.4). As of 2002, upland fields in the devolved forest accounted for 17.5 per cent of the total upland area (90.8ha of 518.6ha) of the surveyed households. Paddy land in the devolved forest was 14.4 per cent of the total paddy land (8.2ha of 57ha). In four villages, local people had roughly 30 per cent or more of their dry land in the devolved forest. By contrast, in two villages around 5 per cent or less of the upland fields was from the devolved forest. As for paddy land, only six out of thirteen villages had paddy land in the devolved forest, but two villages had more than 50 per cent of their paddy land there.

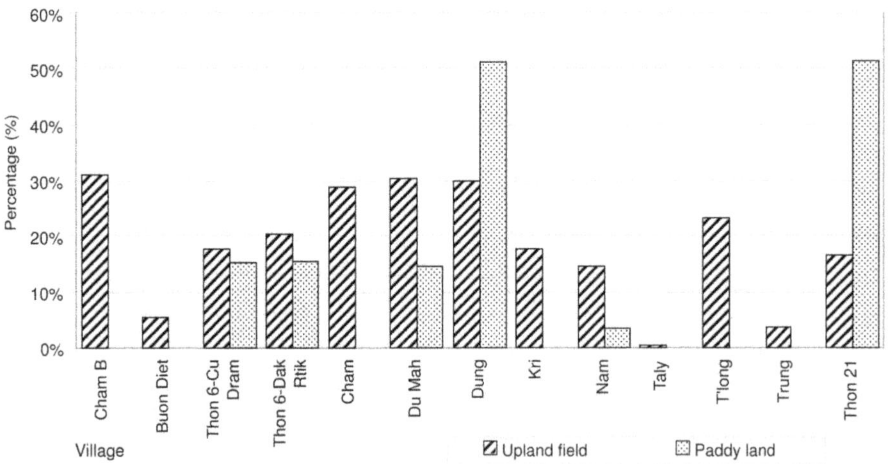

Figure 12.4 *Fields in devolved forests versus total farmed land, 2002*

Food production from forest versus total household food production

Food production from fields in the devolved forest contributed to total household food production (Figure 12.5). On average, 15 per cent of the total household food produced came from the devolved forest in 2001–2002. In one village, this figure was almost 60 per cent. By contrast, in four villages 6 per cent or less of their total food production came from the devolved forest. In the remaining villages, the figure was between 10 and 25 per cent.

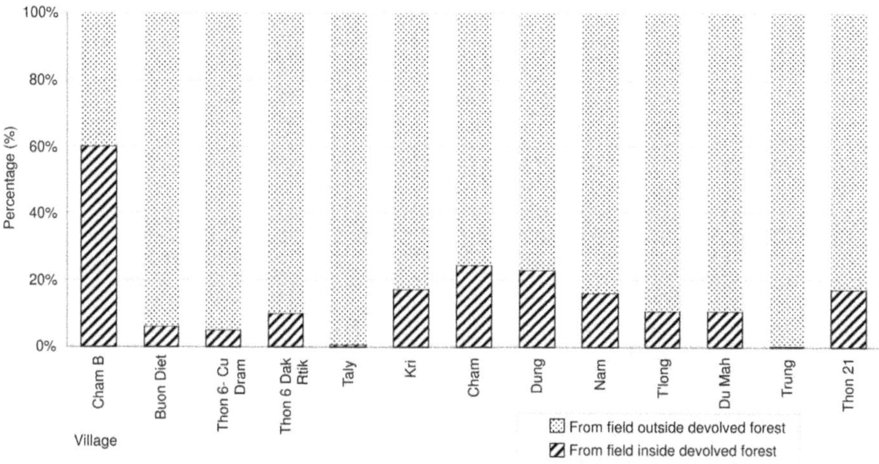

Figure 12.5 *Food crops from forest fields versus total food production*

Compared with 1999, food production from fields in the devolved forest generally increased in the study villages. The share of food produced from devolved forests increased from 7 per cent of total food production in 1999 to 15 per cent in 2001–2002. In absolute terms, the amount of food produced in the devolved forest increased threefold. An increase in the absolute value of food production in the forest was observed in 11 villages. Only in two villages was less food produced in 2002 than in 1999.

Contribution of devolved forest to cash income

Though modest, cash income from devolved forest fields made a contribution to household cash income (Figure 12.6). On average, around 5 per cent of the income from cash cropping came from the fields in the devolved forests. In six villages, no income was observed to come from the forest. However, in one village 48 per cent of the income from cash cropping came from fields in the forest. In other villages, the figure was between 4 and 10 per cent.

Besides direct cash income from crop harvests, new fields in the devolved forest also affected cash income from cropping by making land in other places available for production. When people grew food crops on the new fields in the devolved forest, existing fields could be shifted from subsistence food production to cash cropping.

In general, the average share of cash income from devolved forest fields relative to the total cash income from cropping rose from 1.4 per cent in 1999 to 7 per cent in 2001–2002. In absolute terms, income from fields in the devolved forests in 2001 was approximately 2.5 times that of 1999.

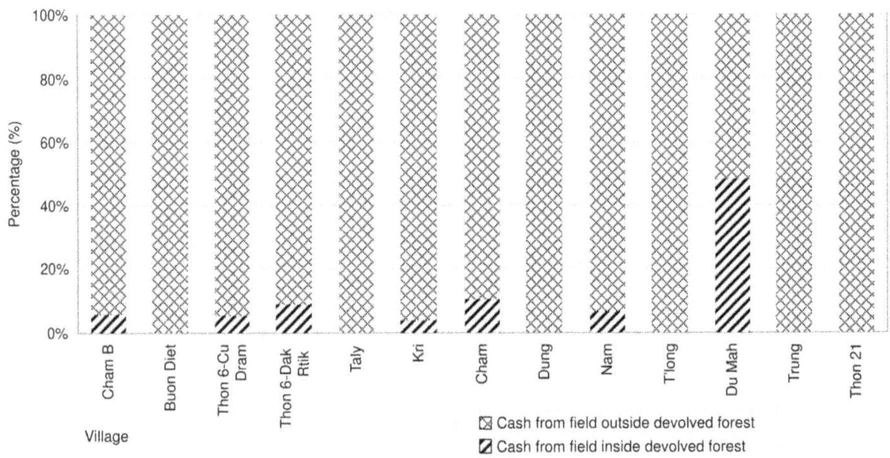

Figure 12.6 *Cash income from forest fields versus total cash from cropping*

Forest devolution and access to credit

The Land Law provides that a forest land-use title can be a valid document for formal credit. Nevertheless, the actual use of forest Red Book certificates as collateral for loans was rather poor. Of the total of 176 households with certificates in the sample, only 6 households in one village, with support from the local authorities, were able to get a loan from a local bank by using their certificates as collateral. One reason was the low market value of the forest and high expected cost of liquidating this asset in case of failure of repayment. A title for agricultural land or residential land, which had also been given to local people, would be more acceptable for mortgaging purposes than a forest land-use title. Furthermore, the complicated procedures for loan applications discouraged farmers from trying to borrow.

Nevertheless, the experience in Dak Lak shows that forest devolution provides a chance for farmers to get credit through a development project. The six households are an example. In cooperation with local authorities, the Dak Lak Department of

Agriculture and Rural Development selected six households to experiment with a farm forest model on their devolved forest. With the pledge of local authorities to back them up in case of default, each household was given a loan of 2 million to 5 million VND, using their forest Red Book certificates as collateral. The loan was then used for cattle raising and cashew plantation.

Household costs related to devolution

In addition to benefits, households receiving forests also had costs to bear. Their most important cost was labour, which was incurred both during the devolution process and after its completion. During the forest devolution process, each household had to attend on average two meetings and to spend a day in the field. Until the completion of forest devolution, a minimum of two working days, valued at 40,000 VND in cash equivalence at current market rates, were spent by each household.

The realization of benefits also involved labour expenses. In theory, forest recipient households were expected to abide by state regulations on the use of forest resources. They had to acquire logging permits for timber and follow the logging plan for large-scale commercial logging. They were also required to seek permission when they wanted to change the land use in the devolved forest, including converting forest land to cropping land. In addition, local forest managers were expected to regularly patrol their forest to detect, stop and report unauthorized uses of the resources.

For a forest recipient household, the greatest challenge was obtaining the required permits for logging timber and converting forest into cultivated land. Even the district officials were unclear about how farmers could get official permission. Farmers had to comply with administrative procedures and travel to the district capital to apply for permits. For example, one farmer trying to apply for a logging permit to build a house made three trips to the district centre 30km away without any success. In the end, he decided to log the timber without a permit, taking the risk of being caught and having his logs and equipment confiscated.

Forest patrolling also required labour from the household. In most villages, forest recipient households said they had visited their forest between 10 and 15 times during the previous year. Most of these trips, however, were combined with other activities, such as collecting forest products.

Labour for opening new fields, logging or collecting non-timber forest products was the most prominent cost, particularly in the absence of modern machinery and equipment. Cutting 1ha of secondary or young forest to create new fields, for example, took 40 days on average. Once the trees dried, it took 5 to 10 days of labour to clear the ground using fire. Approximately 10 to 15 more days of labour were needed to fence the field. Altogether, it generally took 45 to 60 days to convert 1ha of devolved forest into cropland.

Discussion and Recommendations

Forest devolution has the potential to improve local livelihoods, yet the participation of local people in the protection of devolved forests hinges on transparent and workable benefit-sharing mechanisms that pay attention to poor and disadvantaged members of the community. To date, most of the benefits from devolved forests have accrued to a small group of households (Figures 12.7 and 12.8).

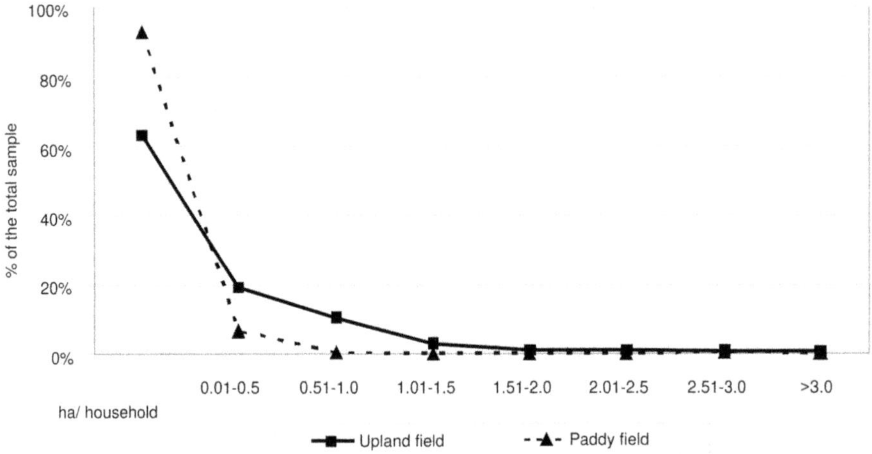

Figure 12.7 *Distribution of forest fields among households*

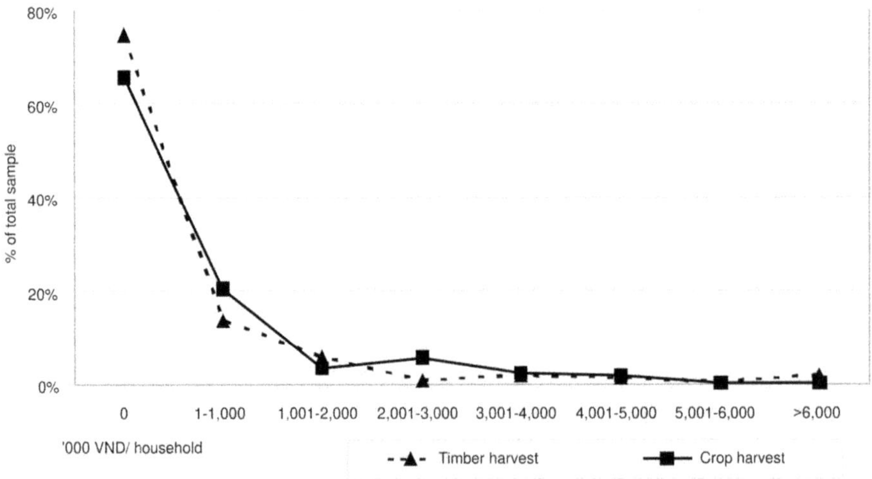

Figure 12.8 *Distribution of crop outputs and timber harvests among households*

For example, 63 per cent of the local households did not have any upland fields in their devolved forest, and 30 per cent had no more than 1ha; only 7 per cent had more than 1ha. The distribution of paddy land was even more uneven. Around 93 per cent of the surveyed households did not have access to paddy land in the devolved forest – all the land belonged to just 7 per cent of the sample.

The distribution of income from harvesting in the devolved forest was also very uneven. Around 75 per cent of the sample did not have benefits from timber and 14 per cent had less than 1 million VND. Around 6 per cent had 1 million to 2 million. By contrast, 2 per cent of the population had more than 6 million.

In most cases, economic benefits accrued to wealthy households and those who had labourers (Figure 12.9). As measured by an index representing ownership of a good house and expensive household assets (for example a motorbike, sofa

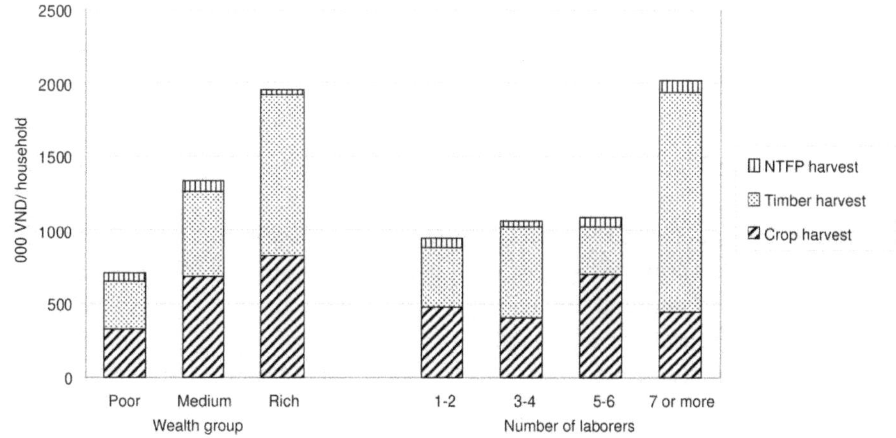

Figure 12.9 *Distribution of forest harvests by income and labour force*

or TV), around 181 poorer households in the sample (56 per cent of the total) enjoyed only 37 per cent of the total value of economic harvests. The same amount of benefits was derived by 96 households of medium wealth class (30 per cent of the total sample). The remaining 26 per cent of the benefits went to the richest 14 per cent of the sample. On average, the benefits from devolved forest derived by an economically medium household and a rich household were around 190 per cent and 280 per cent, respectively, of that of a poor household.

Furthermore, some benefits accrued to those without legal rights to the devolved forest (Figure 12.10). Around 57 per cent of households without forest Red Book certificates in the sample derived some benefits from devolved forest, compared with 76 per cent of the households with certificates. On average, the value of benefits that went to the former was relatively low, at 1.1 million VND, compared with 1.9 million VND for certificate holders. However, households

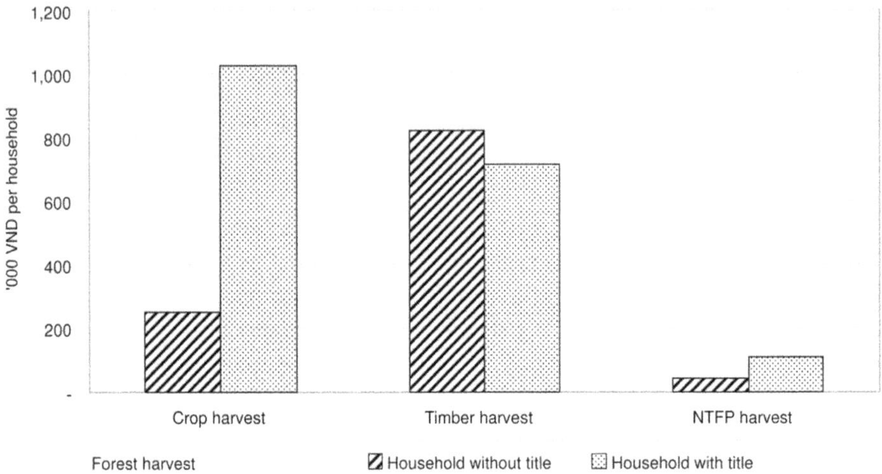

Figure 12.10 *Distribution of forest harvests by legal status*

without title were generally collecting more timber from the forest than those with title.

Although benefits from devolved forest made a significant contribution to the household economy in general, forest recipient households face challenges in protecting the forest and maintaining its benefits in the future. It was difficult for local people to exclude fellow villagers who had been relying on the forest for supplies of materials for daily use. But the use of forest benefits by those without legal rights may undermine the incentive of local forest managers to spend their own labour to protect the devolved forest. Furthermore, since most of the benefits accrued to wealthy or labour-rich households, those without adequate resources to take advantage of the forest may not have sufficient incentive to continue protecting it.

From an economic point of view, the problem is the absence of a transparent and feasible benefit-sharing mechanism and a monitoring system for implementing such a mechanism. For local farmers in Dak Lak, the benefit-sharing mechanism introduced by the forest devolution programme was difficult to comprehend, and it was unclear what organizations would help them in realizing their devolved rights. That a farmer made three trips to the district centre and still failed to obtain a logging permit indicates that the system is not working well.

From a policy point of view, findings from this study have the following implications. First, forest devolution can provide local people with true economic benefits and stimulate their interest in local forest management. Nevertheless, it is important to take into account that devolution benefits are likely to vary among households and villages. Such variations reflect the varying ability of local households and villages to make use of the devolved forest.

Second, the state cannot simply give local citizens the rights to the devolved forest and expect them to benefit. Rights alone are insufficient for acquiring true economic benefits – other resources, such as labour and capital, are also needed. The state should enable people to benefit from forests.

Third, resource-poor households are often in a poor position to benefit from government policies. Rural development policy, therefore, should target or positively discriminate on behalf of these households.

Fourth, the role of external support is important in forest devolution. Devolving the rights to forest to local people is a radical change from Vietnam's long-standing state management in forestry. Such implementation therefore requires significant facilitation from outside, particularly in countries with complicated systems of laws and regulations.

Finally, it is important to develop a clear and feasible monitoring system for state policy at the field level. For forest devolution, such a system should track actual local resource use and make sure that both rights and duties are respected and followed. In addition, a clear, understandable system will help local people avoid unnecessary costs in trying to realize their rights.

Acknowledgements

Funding for field research was from the Tropical Ecology Support Program under the German Agency for Technical Cooperation. The Sustainable Management of Natural Resources in the Lower Mekong Basin Project and later the Watershed Management Component of the Agriculture, Irrigation, and Forestry Program under the Mekong River Commission handled the funding. The project was affiliated with the Dak Lak Department of Agriculture and Rural Development in Vietnam and the Institute of Agricultural Economics and Social Sciences, Humboldt University, Berlin, in Germany.

This chapter benefited greatly from comments by participants from the Workshop on Forest Governance and Decentralization in Asia and the Pacific organized in Yogyakarta, Indonesia, in September 2006. I would also like to thank the Center for International Forestry Research for sponsoring my participation in the workshop and for institutional support in writing this chapter. Thanks also to five anonymous referees who gave invaluable comments on an earlier version.

References

Bui, D. M. (1989) 'Viec dua lao dong dan toc vao nong lam truong [Involving indigenous people as workers in the state forest and agricultural enterprises]' in *Tay nguyen tren duong phat trien [Central highlands on the way to development]*, Nha Xuat Ban Khoa Hoc Xa Hoi (Social Science Publishing House), Hanoi, pp449–466

DARD (Dak Lak Department of Agriculture and Development) (2001a) 'Bao cao ket qua buoc dau thuc hien thi diem giao dat lam nghiep tai tinh Dak Lak trong hai nam 1999–2000 [Report on the initial results of the experimental forest land allocation in Dak Lak in 1999–2000]', DARD, Dak Lak, Vietnam

DARD (2001b) 'Chien luoc phat trien lam nghiep tinh Dak Lak giai doan 2001–2010 [Dak Lak forestry development strategy for 2001–2010]', DARD, Dak Lak, Vietnam

MARD (Ministry of Agriculture and Rural Development) (1999) *National Forum on Forestry*, Agriculture Publishing House, Hanoi

Ngo, T. D. (1989) 'Su phat trien kinh te xa hoi trong cac dan toc it nguoi o Dak Lak [Socio-economic development of ethnic minorities in Dak Lak]', in *Tay nguyen tren duong phat trien [Central Highlands on the Way to Development]*, Nha Xuat Ban Khoa Hoc Xa Hoi (Social Science Publishing House), Hanoi, pp362–376

Nguyen, D. V., To, M. D., Ha, C. C., Trinh, H. D. and Nguyen, K. H. (2001) *Lam nghiep Viet Nam 1945–2000 [Vietnamese Forestry 1945–2000]*, Agricultural Publishing House, Hanoi

Nguyen, T. Q. (2005a) 'Trends in forest ownership, forest resources tenure and institutional arrangements: Are they contributing to better forest management and poverty reduction? The case of Vietnam', report to FAO Forest Resource Assessment, Hanoi

Nguyen, T. Q. (2005b) *What Benefits and for Whom? Effects of Devolution of Forest Management in Dak Lak, Vietnam*, Shaker Verlag, Aachen, Germany

Nguyen, T. D. (2005c) 'Gioi Thieu Ve Du An Trong Moi Nam Trieu Ha Rung [Introduction about the Five Million Hectare Reforestation Program]', Five Million Hectare Reforestation Program Secretariat, Hanoi

Nguyen, T. Q. (2006a) 'Forest devolution in Dak Lak, Vietnam: Improving forest management or livelihoods of the poor?', in S. Mahanty, J. Fox, M. Nurse, P. Stephen and L. McLees (eds) *Hanging in the Balance: Equity in Community-based Natural Resource Management in Asia*, Regional Community Forestry Training Center for Asia and the South Pacific (RECOFTC) and East–West Center, Bangkok and Hawaii, pp200–222

Nguyen, T. Q. (2006b) 'Forest devolution in Vietnam: Differentiation in benefits from forest among local households', *Forest Policy and Economics*, vol 8, pp409–420

Nguyen, V. A. (1989) 'Mot vai van de kinh te – Xa hoi o Dak Lak [Some socioeconomic problems in Dak Lak]', in *Tay nguyen tren duong phat trien [Central Highlands on the Way to Development]*, Nha Xuat Ban Khoa Hoc Xa Hoi (Social Science Publishing House), Hanoi, pp45–54

Nguyen, X. V. (2003) 'Giao rung tu nhien tai Dak Lak: Co hoi va thach thuc [Allocation of natural forest in Dak Lak: Opportunities and challenges]', Dak Lak Department of Agriculture and Rural Development, Dak Lak, Vietnam

Roth, P. (2005) 'Conceptual framework for community-based forest development in Quang Binh: Sustainable management of natural resources in central Vietnam', Sustainable Management of Natural Resources in Central Vietnam (SMNR-CV) Project, Quang Binh, Vietnam

Sikor, T., Nguyen, T. Q. and Tran, T. N. (2005) 'Learning from devolution: An assessment tool developed in Vietnam for the Forest Department', *International Forestry Review*, vol 7, pp320–331

Tran, T. N. (2005) *From Legal Acts to Village Institutions and Forest Use Practices: Effects of Devolution in the Central Highlands of Vietnam*, Shaker Verlag, Aachen, Germany

Tran, T. N., Nguyen, T. Q. and Sikor, T. (2003a) 'Local impact assessment after forest land allocation: Manual', Dak Lak Department of Agriculture and Rural Development, Dak Lak, Vietnam

Tran, T. N., Nguyen, T. Q. and Sikor, T. (2003b) 'Local outcomes of forest land allocation: Evidence from Dak Lak', Dak Lak Department of Agriculture and Rural Development, Dak Lak, Vietnam

Tran, T. N., Nguyen, T. Q. and Sikor, T. (2004) 'Local impact assessment after forest land allocation in Dak Lak, central highlands of Vietnam', Deutsche Gesellschaft für Technische Zusammenarbeit (GTZ) GmbH, Eschborn, Germany

Tran, T. N. and Sikor, T. (2006) 'From legal acts to actual powers: Devolution and property rights in the central highlands of Vietnam', *Forest Policy and Economics*, vol 8, pp397–408

Wode, B. (2003) 'Report on community based forest management concept', Rural Development Project, Dak Lak, Vietnam

13

Decentralization and Forest Governance in Asia and the Pacific: Trends, Lessons and Continuing Challenges

Doris Capistrano

Forestry decentralization in the Asia-Pacific region continues to be a work in progress. Different countries have taken varied paths and have opted for different types of decentralization (Ferguson and Chandrasekharan, 2005). As in other parts of the world, countries in the Asia-Pacific region continue to experiment with the degree and pace of decentralization and to negotiate the balance of power between central and lower levels of government and among stakeholder groups (Enters et al, 2000; Brown and Durst, 2003; Colfer and Capistrano, 2005; FAO 2006). Some countries (for example China, Vietnam, Lao PDR, New Zealand and Indonesia) have embarked on relatively rapid and sweeping reforms to effect decentralization. Such radical reforms have typically been ushered in by political discontinuities, such as the fall of the Suharto regime in Indonesia, the election of a new government in New Zealand, and rapidly developing market opportunities in China, Vietnam and Lao PDR (Box 3.2 and Chapters 6, 8, 10 and 12 of this volume).

Most countries in the region, however, have opted for gradual, incremental decentralization. Even Indonesia, which started with a bang, has downshifted to a more cautious pace as the initial exuberance with decentralization increased conflicts and institutional complexities (Siswanto and Wardojo, 2005; Barr et al, 2006). As a political process, decentralization necessarily entails continuing negotiation of power and authority over forests and related resources. As a social process, decentralization also involves continuing struggles over ideology and guiding philosophy and over the alternative meanings and directions they imply.

In contrast to the heady early days of decentralization three decades ago, the types of decentralization implemented during the past 15 years have tended to be more pragmatic and give more explicit attention to equity, livelihoods and market opportunities. There also appears to be greater appreciation for the long-term nature of the process of decentralization and growing recognition that, inevitably, decentralization requires continual learning, adaptation, and technical, administrative and financial support. New skills and capacities need to be developed to enable actors and stakeholders at all levels – communities, local governments, support organizations, state and corporate entities, and national governments – to discharge their roles and responsibilities and take advantage of new opportunities created by decentralization. At the same time, the outcomes from similar decentralization approaches and interventions vary in different countries and locations. Local historical antecedents, the nature and characteristics of institutions, the capacities of key actors, socioeconomic and political dynamics, and options for the use of forests and related resources all affect the end results (Chapter 3 of this volume; Capistrano and Colfer, 2005).

Backlash from dominant stakeholders and entrenched interest groups protesting the erosion of their power and working to recapture lost ground has caused the pendulum in some countries (for example Nepal, Indonesia and the Philippines) to swing back towards greater central control in recent years (Barr et al, 2006; Chapters 5 and 11 of this volume). In many countries, the opposed forces for centralization and decentralization play out in different decision-making forums. These counter-pressures can lead to periodic cycles of decentralization and recentralization, observed in many countries around the world (Capistrano and Colfer, 2005). At times these countercurrents become manifest in contradictory policies, processes or mechanisms that negate each other. This is akin to giving with one hand while taking with the other to maintain an uneasy balance or return to a pre-decentralization status quo. Examples include Vietnam's and Lao PDR's state devolution of collective forest management to individual households and companies while designating much of the devolved forests as nature reserves and protected areas (Chapters 8 and 12 of this volume).

Lessons from countries that have already undergone decentralization and devolution indicate that central control and decentralization need to be complementary and are perhaps best viewed as two sides of the same coin. Without a strong and effective central government, decentralization can lead to loss of policy coherence, heightened inequities, accelerated fragmentation and forest loss (Box 3.2 and Chapter 2 of this volume). It can also weaken integrated land-use planning, forest firefighting and regulatory enforcement and, under certain conditions, encourage corruption. Analysts have noted that, although some decentralization reforms can be associated with greater state responsiveness and increased levels of citizens' activity, the majority are not. And there are concerns that localization of politics may obscure wider patterns of injustice and undermine collective responses to them (Larson and Ribot, 2005; Ribot, 2005).

As in other parts of the world, in recent decades the overall discourse on decentralization in Asia and the Pacific has become increasingly nuanced, and its instruments more sophisticated and increasingly market-oriented. From an initial perception of decentralization as a somewhat straightforward win/lose allocation of authority between central and local governments and actors, over the years decentralization has become increasingly viewed as a reconfiguration of the bases of power and authority in the context of multi-stakeholder negotiation (Glück et al, 2005). In the redefinition of the relative roles and authority of different stakeholders, mutually beneficial gains can be possible.

Although previous notions of decentralization as top–down transfer of authority remain strong, particularly among state agencies and forestry bureaucracies, decentralization is increasingly understood as a first and necessary step towards participatory forest governance. Decentralization, which has meant privatization in countries such as New Zealand, China, Vietnam and Lao PDR, has also been regarded as a means of capturing greater efficiencies from liberalizing markets. Others see decentralization and the devolution of power and authority that is supposed to follow in its wake as means towards greater democratization. However, to lead to democratization, decentralization must be downwardly accountable and create spaces for genuine, popular participation (Larson and Ribot, 2005).

Earlier rounds of decentralization in the Asia-Pacific region were implemented by governments generally under pressure and with funding from external donors and international development agencies. Reforms tended to focus more on administrative aspects than on the political dimensions of devolution and power sharing. In general, the economic, fiscal and financial dimensions of decentralization tended to be ignored or sidelined in these discussions. As countries in the region look back on their collective experience with the politics and administration of decentralization and devolution, and as external funds to support these processes dwindle, attention has increasingly turned to the requisites for economically and technically viable decentralization. To be viable and sustainable, decentralized systems of forest management will have to demonstrate increased productivity and offer significant returns and long-term incentives. At the same time, those concerned with the environment would like to see decentralization lead to enhanced conservation and environmental improvements over time (Sayer and Maginnis, 2005).

Indeed, the goalposts and standards for assessing decentralization results in the region have been changing along with politics, markets, and understanding of ecology and development. Rising demands and expectations for decentralization, the growing diversity of interests, and the complexity of the politics involved in the process require different, more effective ways of making decentralization work. In reviewing the experience of different countries, it is clear that proponents of decentralization must address persistent problems with participation, equity, tenure, rights, livelihoods and capacities. In the meantime, some old and new areas of concern have emerged: conservation, finance, systems of incentives, corruption and illegal activities in decentralized forest governance.

The authors in this volume and participants in the Workshop on Decentralization in Asia and the Pacific discussed these issues in plenary and working group sessions. Several priority areas for reform were identified, along with recommendations for specific actions. The main recommendations from the workshop are summarized in Box 13.1.

Patterns, Lessons and Continuing Challenges

Frameworks, roles and implementation processes

The cases in this volume reveal major gaps between decentralization policies and their implementation. Intent is not necessarily matched by political will and commitment (Chapter 11 of this volume). Even progressive policies (for example the Philippines' Indigenous Peoples' Rights Act) tend to remain largely on paper (Lynch, 2005; Saway and Mirasol, 2005). Strong demand and credible pressure from below have been required in many cases before states implement decentralization policies. Consistent pressure and show of collective strength have generally accompanied cases of real devolution of power and authority (Shrestha, 2000; Ritchie and Haggith, 2005; Durst et al, 2005; Colchester et al, 2006; Chapter 11 of this volume).

Guidelines and mechanisms for decentralization policy implementation are often unclear or non-existent. Resistance or sabotage by those who stand to lose, including traditional power brokers, competing forest claimants and segments of the state bureaucracy, can derail the implementation of even the best-intentioned policies.

Much remains to be done in clarifying the legal and regulatory framework for decentralization of forest management in Asia and the Pacific. As in other regions of the world, there is often a disjuncture between roles, responsibilities and rewards for forest management (Chapter 3 of this volume; Capistrano and Colfer, 2005; Küchli and Blaser, 2005). And in many countries, forest users and managers still have only a vague understanding of their roles, responsibilities and rights (FAO, 2006). The cases in this volume illustrate the continuing evolution of the definition, allocation and modes of implementation of roles, rights and responsibilities among resource users, managers and stakeholders.

As parties gain experience in their new roles, they seek to address aspects that are unclear, inconsistent, impractical or iniquitous through renegotiations and further reforms. For example, in the 1990s the national and provincial governments of Australia developed and adopted regional forest agreements as a way of addressing their tensions and disputes over forest conservation and trade (Ferguson and Chandrasekharan, 2005). Workable systems of checks and balance are needed to implement decentralization. Clear and accountable frameworks for implementation, monitoring and regulation are important. And the participation of civil society and key stakeholders is essential in defining and achieving the appropriate balance.

Uncertain legal and regulatory frameworks for decentralization create insecurities in entitlements and benefits that undermine the viability and sustainability of local forest management. Unilateral changes in policies, such as those regarding forest product harvest, transport, sales, taxation and rights transfers, affect returns from decentralized forest management. Such unilateral and arbitrary changes weaken confidence and trust in the system and raise doubts about its stability and prospects for realizing expected returns. Unfortunately, such uncertainties are not uncommon.

In the Philippines, use rights of community forest management agreement holders have been suspended or cancelled at least four times since 1999 (Chapter 11 of this volume). In China, national policy on the ownership and management of forest lands changed four times in the quarter-century before 1978, and local administration changed even more frequently. Despite recent reforms, a degree of uncertainty remains among China's tree-farming households. Harvest and shipping permits for timber can change from season to season and be subject to unpredictable increases. Forest taxes have risen in some cases from negligible to 35 to 60 per cent of timber revenues. Even when official state policies are favourable, the arbitrariness of some local administrators' oversight and dealings with tree-growing households also add to the uncertainty at the local level (Hyde et al, 2003).

The cases in this volume show that adequate and honest consultation, transparent and accountable decision-making, and a genuinely participatory approach to decentralization also continue to be major gaps across countries in the region. Clearly, these need to be addressed for viable decentralization in the long term. Processes for participation and procedures for recognition of collective and individual rights in many countries are confusing, cumbersome and in need of simplification. Even in Papua New Guinea, where 97 per cent of the land and forests is owned by communities under customary tenure, forest owners are not necessarily in a position to make decisions or even meaningfully participate in decisions about their forests. The government acquires the rights to harvest timber and develop the condition of forests and confers these rights through public tender to (mostly foreign) logging companies for periods of 5 to 20 years. A cumbersome, 34-step process of securing timber harvest permits involves prohibitively high transaction costs that deter communities from legally harvesting their own timber (Potter and Bun, Box 3.1, this volume).

Lack of trust between different levels of government and among communities and other stakeholder groups also inhibits progress towards effective decentralization and constrains innovation to find locally appropriate solutions to vexing issues. Concerned institutions and forest managers at different levels have to develop and agree on appropriate standards, systems and processes to make decentralization work in their specific contexts. The development of needed skills, capacities and infrastructure remains critically important and requires continuing support. Also required are the resources and enabling environment that can encourage local adaptation and organization that may be more appropriate than centrally prescribed institutional arrangements and forest management plans (Colfer, 2005).

However, more fundamentally, the long-term viability of forest decentralization will require real changes in processes and 'rules of the game', especially on how rights, roles, responsibilities and rewards are allocated and by whom. The rhetoric of participation will have to be translated into real spaces for genuine stakeholder participation. Forest governance reforms are needed to improve the rules of the game and encourage shared decision-making and greater equity in the distribution of benefits and burdens that result. Without needed reforms (see Box 13.1), decentralization is simply another way for central governments to get local people to bear the costs of state forest management (Edmunds and Wollenberg, 2003).

Tenure, rights and privatization

The latest FAO (2006) assessment of the status of forest ownership and tenure in Asia and the Pacific shows that decentralization and devolution in the forestry sector remain limited. The area of forests managed under various decentralized schemes also continues to be limited. Forests in the region are overwhelmingly (86 per cent) public forests,[1] and almost 80 per cent are still under the direct control of central governments. Devolution to local communities involves no more than 10 per cent of forests (18 per cent if small-scale forest holders are included). In general, rights are devolved for degraded forests while well-stocked and valuable forests remain under central control. The FAO study cites evidence of links between clear, secure tenure arrangements and better forest management and more sustainable forest-based livelihoods. However, tenure security alone is not sufficient; it needs to be supported by effective capacity-building and necessary reforms.

Privatization of state forests is among the most significant trends in decentralization in the Asia-Pacific region (FAO, 2006). This trend is most pronounced in China and in centrally planned economies transitioning to liberal market economies, including Vietnam and Lao PDR. In Malaysia, long-term arrangements of 100 years granted to private companies signal a stronger commitment to privatization.

No country's forest policy has changed more dramatically than China's, and many of the changes here are part of a broader trend toward more decentralized management in all sectors of the economy. By 1998, households managed 52 per cent of China's total forest area, including 74 per cent of forest plantations, 92 per cent of economic forests[2] and 93 per cent of bamboo forests (Xu and Hyde, 2005).

Like China, Vietnam has also embarked on a path to privatization.[3] Until the early 1990s, Vietnam's forests were managed solely by state agencies (Nguyen, 2005). This changed dramatically in 1991, when the Forest Protection and Development Law introduced private property rights in forests and established a legal basis for setting up management boards as a new form of state management of protection forests and forests designated for special uses. As in China and other countries in transition from centrally planned to market-based economies, this opening unleashed a range of creative approaches to tenure reforms. There are now a variety of tenure regimes over forests and forest products based on different ways of packaging or bundling the nature and degrees of rights and entitlements over land, resources within the forest over time and space, and flows of services and benefits from the forest ecosystem. At the same time that the rights and entitlements for individual households, communities and corporate entities have expanded, new markets have developed where some of these rights and entitlements can be exchanged. Through engagement in these markets, governments at different levels can more finely tune incentives and disincentives for decentralized forest management.

For countries that have opted to implement decentralization as privatization, New Zealand's earlier experience can be instructive. Between 1980 and 1996, foreign ownership of New Zealand's forest plantations increased from 2 to 48 per cent. Such a dramatic shift in forest resource ownership has implications for governance and regulatory enforcement. With decentralization it is even more important to clarify which level or agency of government is responsible for regulation, enforcement and conflict resolution. In poor countries with weak capacity, regulatory enforcement can be a major issue, particularly when external parties are big and powerful (see Box 3.2).

Indeed, with the move towards greater privatization and market orientation within forest decentralization, clarity, security and protection of rights become even more important. Rights need protection not only from foreign business interests but also from local actors, including the state and its agencies. In Lao PDR, for example, the state designated more than 90 per cent of the village territory it had allocated for community management as protection and conservation forests, in effect curtailing communities' rights to a major portion of the forest products (Lao PDR, 2005; Chapter 8 of this volume). Similarly, in China, though their rights are recognized, individual households have generally not been consulted about major government decisions, such as logging bans or creation of protected areas (Chapter 6 of this volume).

Disagreements over who has rights and who should control forests and forest lands are often at the root of conflicts and destructive forest practices. Decentralization can add layers of complexity and exacerbate disagreements, particularly if the new laws and regulations are conflicting. Decentralization can also pit the interests of local governments against those of the central government with respect to regulating private industry, especially if local governments have incentives to exploit resources faster for revenue generation.

In Indonesia, different interpretations of the definition and location of the country's forests and forest zones, and jurisdiction over them, are at the heart of policy conflicts over public versus private rights and management priorities

(Contreras-Hermosilla and Fay, 2005). Questions about legal jurisdiction and the authority to grant or extinguish rights and issue permits are crucially important in determining which activities are legal or illegal. Local people can be ensnared in webs of illegality, especially when laws and regulations are inconsistent or subject to conflicting interpretations (Resosudarmo and Dermawan, 2002; Lynch and Harwell, 2002; Tacconi et al, 2003; Colchester et al, 2006).

Even when rights are clear, lack of capacity can prevent rights holders from claiming or exercising their rights. In India, for example, the law confers management rights over forests and other natural resources to village governments (*panchayats*). Yet lack of skills, funds, technology, market access and other requisite inputs has prevented many *panchayats* from exercising these rights and discharging their roles (Saigal, 2003).

Forest rehabilitation, fire and conservation

Forest management by communities and households has been credited with rehabilitating degraded lands and transforming landscapes in many countries in the region (FAO, 2006). India's joint forest management, a system of co-management between state governments and village-based forest groups covering almost 17.4 million hectares and involving almost 85,000 villages, is among the few examples of large-scale decentralization (Bahuguna et al, 2004). Although joint forest management has had mixed results in terms of livelihoods and equity, it has contributed to the rehabilitation and re-greening of India's degraded forests and wastelands (Saigal et al, 2005; Singh and Varalakshmi, 1998). Detailed ecological studies show general improvements in tree cover, crown density and forest productivity in joint forest management areas (Ravindranath et al, 2004). For example, between 1995 and 1998, tree density in nine randomly selected sites in the Eastern Ghats showed improvements ranging from 139 to 704 per cent. An analysis of indicators of ecological change in 5 per cent of all joint management groups in the state of Andhra Pradesh over the same period revealed similar improvements. Growing stock increased by 50 per cent; increases in non-timber forest products ranged from 26 to 165 per cent; overall regeneration ranged from 13 to 145 per cent; increase in species diversity ranged from 29 to 165 per cent; and water tables registered an average rise of 3.6m (Rangachari and Mukherji, 2000; Banerjee, 2004).

In China, despite the limitation on timber markets, the effects of privatization on forest cover were rapid and substantial on both agricultural and forest lands. In its windy and erosion-prone North Central Plains, forest cover expanded from about 5 per cent in 1977 to 11 per cent by 1988. Improvement in property rights was also determined to have positive impacts on production and conservation. Conservation investment in trees (a response to improved property rights) explained 5 per cent of the increase in agricultural output from 1977 to 1984 and 20 per cent thereafter, as newly planted seedlings matured into larger trees and formed better protection against eroding and desiccating winds (Yin, 2003).

Preventing and fighting forest fires can be more complicated under decentralized forest management. In South Korea, fire prevention and control systems needed to be restructured to better complement the decentralized structure of the forest management bureaucracy (Chapter 7 of this volume). The involvement of villages and local forest users in forest fire prevention and firefighting has resulted in fewer incidents of fire in community-managed forests in India (Bahuguna et al, 2004). In Indonesia, many forest fires are set for land clearing and illegal forest land conversions to crop plantations (Barber and Scheweithelm, 2000). In theory, decentralization and the improved forest governance that is expected to ensue should help reduce forest fires. This has not been the case in Indonesia, where fires continue to affect large tracts. In China, fragmentation of land under private household management has raised the cost of forest protection. Boundaries per unit of forest land are relatively longer in small plots than in larger plots. The costs of boundary markings, patrolling and protection from fires and illegal activities have increased because of land fragmentation under decentralization through privatization. This has limited the ability of many households to effectively protect their forests from fire and theft.

Overall, patchy information and lack of data make it difficult to assert general conclusions about the links between decentralized forest management, forest land rehabilitation and fire protection. The impacts of decentralization on fire prevention vary in different contexts. Links between decentralized forest management and biodiversity conservation also remain uncertain, although some have argued that protected areas are better retained under central control (Sayer et al, 2005; Sayer and Macginnis, 2005).

What is clear, however, is that the momentum for local forest management under different types of decentralization schemes, and any environmental benefits they provide, may be difficult to sustain. In many cases, there is little incentive for local governments, households and communities to continue to engage with decentralized forest management schemes (Mayers and Bass, 1999). Particularly where the initial quality of forest resources is poor, where infrastructure, capacities and inputs are deficient and markets underdeveloped, the burdens of local forest management can outweigh the benefits (Capistrano and Colfer, 2005).

Corruption and illegal forest activities

Corruption, illegal forest activities and problems with law enforcement are common, though their nature and extent vary among countries in Asia and the Pacific. Corruption and illegal forest activities can be a significant factor in increased forest-related violence and civil unrest (Peluso and Watts, 2001; Tacconi et al, 2003). They can also be so pervasive in some countries that progress to address them will require concerted reforms and interventions beyond the forest sector alone (FAO/ITTO, 2005).

Decentralization has been expected to serve as a counterweight to organized corruption, state and national elite capture of forest benefits, and collusion by corrupt and inefficient forestry bureaucracies (Chapters 3 and 11 of this volume). It is expected to provide local stakeholders and communities with greater power and control over local forest resources, strengthen incentives for protection from illegal logging and unauthorized resource use, and foster greater self-regulation among local users. But decentralization can be a double-edged sword. It presents both threats and opportunities for addressing corruption, illegality and criminality.

Under certain conditions, effective collective action and decentralized forest governance have resulted in more effective regulation and forest law enforcement. Reductions in illegal encroachment and forest land conversion, illegal logging and wildlife poaching, and corruption, for example, have been reported in community-managed forests in South Asia (Rangachari and Mukherji, 2000; Bahuguna et al, 2004). In these cases, individual local government officials, local actors and institutions are often crucial in determining outcomes. Commonly understood rules and consistency of their enforcement have been identified in detailed studies as an important factor (Gibson et al, 2000; Ostrom et al, 2002). Successful enforcement at the local level depends in part on individuals who generally agree on what rules they should follow and why these rules have been adopted. Without this agreement, there is less incentive to comply. Vigorous efforts to enforce the rules can result in bribery and corruption or conflict and violence (Gibson et al, 2000).

Decentralization can also weaken regulatory enforcement and make the situation more complex and difficult. When centrally organized mechanisms for rent-seeking fall apart and fragment into competing factions all extorting rent, decentralization can end up decentralizing corruption as well. This phenomenon appears to have occurred, for example, in some areas in Indonesia and the Philippines (Siswanto and Wardojo, 2005; Barr et al, 2006; Chapter 11 of this volume).

Inconsistencies in decentralization laws, policies and regulations can exacerbate policy confusion and create grey areas in which corruption tends to flourish. Uncertainties during the process of transition to decentralized management and lack of clarity on the locus of authority for granting permits and sanctions provide fertile ground for corruption and illegal activities. Overly burdensome bureaucratic requirements, coupled with high transaction costs of compliance, can also make it difficult for poor, small-scale forest users to comply (Potter and Bun, Box 3.1, this volume; Tacconi et al, 2003; Obidzinski et al, 2006). Unrealistic prohibitions, such as indiscriminate logging bans or prohibitions on use of forest resources without providing alternatives can render customary forest uses for subsistence and livelihoods technically illegal (Colchester, 2006). In cases of violations, local forest users and small-scale loggers tend to be penalized more severely, while large-scale illegal loggers, financiers and powerful perpetrators of organized forest crimes tend to escape sanctions (Setiono and Husein, 2005). Clear and consistent laws and policies and fair and just regulations and their consistent enforcement are essential foundations for any long-term approach to addressing corruption, illegalities and forest crimes.

Alliances with civil society and use of the media to raise public awareness and generate popular support, advocacy and direct action have been employed with varying degrees of effectiveness to address forest-related corruption and illegal activities. In the Philippines in the mid-1980s, an alliance of civil society organizations, anti-Marcos coalitions and religious groups found political space and started advocating for reforms to clean up widespread corruption and redress inequities. Their advocacy led to the cancellation of logging concessions to Marcos's cronies, which had been lucrative vehicles for environmental plunder and subsidies to a privileged few. This resulted in the reduction in the number of large-scale logging concessions (Chapter 11 of this volume). Some of the land was subsequently reallocated to various forms of local community and household management. Advocacy by these allied stakeholder groups and popular support for reforms eventually led to a reconfiguration of the country's policies, with far-reaching implications for forestry and related sectors.

Addressing corruption, criminality and illegality is fraught with risks and requires a variety of approaches at different levels to be effective. Several tactics have been used singly or in combination. These include media pressure, direct action such as 'naming and shaming' perpetrators, economic sanctions such as freezing assets or 'bankrupting the corruptors', local community empowerment to apply 'pressure from below', and international-level cooperation in trade, finance, investment regulation and law enforcement.

Breaking the links between corrupt business interests and state decision-makers is essential in the fight against corruption. Workshop participants agreed that under decentralized forest management, approaches need to be adapted to specific situations in different provinces, districts and locations. Results are expected to vary greatly depending on the capacity and commitment of local government, local institutions and other actors. There was also a common expectation that central government agencies play a strong, clean oversight role. Appropriately targeted international pressure and support from donor governments, trading partners, international agencies, global civil society and advocacy organizations can help strengthen the resolve and commitment of governments reluctant to tackle corruption, illegal logging and forest-related crimes. International initiatives such as Forest Law Enforcement Governance and Trade (FLEG/T), the Asia-Pacific Forestry Commission, and partnership forums such as the Asia Forest Partnership can also help encourage and facilitate necessary action.

Gender, equity and external alliances

Decentralization is a social process embedded in local politics, social relations and histories, which shape it and are in turn shaped by it. Patterns of local hierarchy, dominance, exclusion and inequities are woven into these contexts and form the subtext for decentralization in specific countries and locations (Chapter 4 of this volume). Inequity – within and among geographic regions, gender groups, ethnic and tribal groups, castes, economic classes, communities, and stakeholder groups – is typically an element of this context. Unfortunately, inequity persists as one of the major issues in forest decentralization in the Asia-Pacific region as elsewhere.

Uneven access to resources and opportunities to participate in forest-related decision-making limits the effectiveness and unfairly distributes benefits and burdens of decentralized forest governance. Poor and low-status households, regions, ethnic minorities, forest-dependent communities and women bear a disproportionate share of the burdens and share least in the benefits (Carter and Gronow, 2005; Colfer, 2005). They also tend to have the least influence in decision-making but are the most affected by insecurities wrought by decentralization, especially on vital issues such as forest land tenure, rights of resource access and use, and benefit sharing. The chronically poor are especially excluded from access to institutions even at the most local level. Even those who engage actively often find participation an onerous activity that detracts from their pursuit of livelihoods and helps perpetuate the status quo (Cleaver, 2005).

Across countries in the region, women and disadvantaged groups are typically under-represented in institutions, organizations and forums for forest-related decision-making. Village forest committees, state forest departments and different levels of government agencies dealing with forests are generally dominated and controlled by men, elite groups and privileged classes. This domination and control of processes for representation, agenda setting, modes of operation and resource allocation inhibit the participation of women and disadvantaged groups. Interventions and changes in ways of working are necessary to enable women and marginalized groups to participate and benefit from their participation (Chapters 3 and 4 of this volume). These include targeted capacity-building, including the capacity to negotiate; upgrading of communication, literacy and educational skills; increased access to credit and other livelihood resources; and adjustments in the conduct and scheduling of activities.

External actors have been influential in shaping local contexts and forest-related interactions. The actions and decisions of foreign donors and international environmental and civil society organizations, as well as central and state government agencies in distant locations, are increasingly affecting local people and their relations to forests and other actors. Depending on who they are and what their interests are, external actors can facilitate flows of funds; open new channels of access to information, technical resources and inputs; broker and bridge contacts with other external actors and institutions; introduce new practices and standards of behaviour; and catalyse the creation of new organizations and institutional arrangements. In the process, however, they may deliberately or inadvertently also create new elites and alternative bases of local power and legitimacy.

External interventions, such as donor-supported projects and contractual arrangements with state and private corporate entities, can both mitigate or exacerbate local inequities and introduce new norms and standards of behaviour. Externally mandated reforms in the composition and functioning of local organizations and institutions have created some openings for women and disadvantaged groups in some countries. Reserving a minimum number of village forest council seats for women and low-caste representatives in India and Nepal, for example, helped boost women's participation in village resource management. The space for women's participation has gradually expanded as increasing numbers of women in village-based forest councils contest and get elected to positions in local government (Boaz and Boaz, 2006; Chapter 5 of this volume).

When needed capacity, confidence-building and space for participation are not adequately provided, however, women and disadvantaged groups can become mere tokens. Their presence and marginal participation can lend legitimacy to captured institutions and organizations with few benefits for themselves and other disadvantaged groups. Rather than challenging the dominance of entrenched interests, in many cases, women and disadvantaged groups have opted out of male-dominated and elite-captured organizations and institutions. They have tended instead to establish their own organizations and design their own processes. They have also chosen to frame and address forest-related issues within the context of local development and their own needs for livelihoods and well-being. Women have generally fared better and derived greater benefits from their investment in time and energy when they control the terms of their participation and have greater latitude in framing issues and deciding how best to address them (Chapter 4 of this volume).

New capacities and alliances among the disempowered and marginalized are critical for decentralization to deliver real changes in the rules of the game and create genuine spaces for voices from below. Some federations and umbrella organizations of forest users who on their own would be powerless have employed this strategy. Advocacy has been an essential strategic element.

Effective advocacy is a necessary part of the struggle to promote particular agendas, preferred interpretations and courses of action in the public policy sphere. Successful advocacy strategies have been typically accompanied by building alliances with internal and external actors, building organizations, mobilizing members and supporters, and staging marches and other activities to demonstrate collective strength (Ritchie and Haggith, 2005; Boaz and Boaz, 2006; Chapters 4 and 11 of this volume).

Nepal's national federation of community forest user groups has successfully used this strategy with financial, technical and moral support from its donors and allies. This helped make the Federation of Community Forest User Groups a significant player in Nepal's forest policy discussions and enabled it to play a role in the country's recent transition to democracy. The strategy required organizational strengthening and capacity-building within its ranks, which also helped reduce inequities and corruption among members of the federation. Despite their progress, however, elite members continue to capture a disproportionate share of the benefits from the federation's forest management efforts. Domination by powerful groups at both the community level and within the federation also continues to be an issue of concern (Chapter 5 of this volume).

Financing decentralization

Pressure from donor governments and international development agencies has been a driving force for decentralization in many countries in the Asia-Pacific region. Decentralization was prescribed as a way of downsizing inefficient bureaucracies, off-loading responsibilities for degraded or low-value forest land, and reducing government deficits as conditions for structural adjustment lending in the 1970s and 1980s (Capistrano, 1994; Larson and Ribot, 2005; Ribot, 2005). Because decentralization in most countries came about partly in response to tight budgets and heavy debt burdens, however, funding for decentralization has generally been inadequate. Funds for decentralized forest management have come largely from central and local governments, local communities, external donors and the private sector. Of these, the private sector remains a relatively minor source of funding for decentralized forest management; its potential as a funding source remains largely untapped.

Decentralization efforts in most countries in Asia and the Pacific continue to rely to varying degrees on external donor funding. In India, foreign donors provided roughly 30 to 35 per cent of the total budget of the forestry sector between 1981/1982 and 1998/99 (Saigal, 2003). The share of donor funding for forestry in some countries is even higher. During the past decade, a major portion of donors' forestry funding tended to go to plantation development, rehabilitation of degraded forest lands, protection forests and conservation reserves. Improved decentralization, community forestry, and livelihoods and governance reforms have received less funding despite the high profile and publicity given to these topics in national and international forestry discussions.

In countries that depend mostly on external funds, the future direction and sustainability of decentralization are subject to the vagaries of fluctuating donor budgets and changing priorities. The need to show measurable success within short timeframes and to demonstrate impacts against rapidly changing donor benchmarks complicate decentralization planning and implementation. Changing standards of achievement as preconditions for continued support make it difficult to sustain decentralization initiatives, even in cases where real positive results are being achieved. Some factors, such as international trade, foreign debt and currency fluctuations, are likely to significantly affect decentralization outcomes. Yet they frequently lie beyond a country's control and pose continuing uncertainties (see Box 3.1).

Future funding to sustain decentralization and ensure its long-term viability will eventually have to be generated by the countries themselves. This will require putting in place incentives that will make decentralized forest management schemes attractive options for households, communities, the private sector and different levels of government. It will also require sufficiently strong political support and commitment from governments to commit funds to make decentralization work.

Incentives and investments

In practice, decentralization has generally meant devolving the costs and burdens of forest regeneration, protection and management to local governments, communities and households, with little authority and uncertain benefits (Capistrano and Colfer, 2005). The most productive and valuable forests in Asia and the Pacific and the rights to these resources remain largely centrally controlled. There is also a tendency to undervalue and under-compensate communities' and local peoples' investments and contributions in forest management while giving preferential treatment to investments by external actors and corporations (White and Martin, 2002; Molnar et al, 2004; Scherr et al, 2004; Box 3.1 and Chapter 11 of this volume).

Different actors and stakeholders have different interests and respond to different incentives to engage in decentralized forest management schemes. Secure tenure and an expanded set of rights and entitlements to land and forest products are generally important incentives for communities and households. Early decentralization efforts were designed based on this premise. They were structured to give communities and households access to (mostly degraded) forest lands to enable them to grow subsistence crops and collect forest products. In exchange, communities and households agreed to assume responsibility for forest protection and forest land rehabilitation. But as the actual costs and returns from forestry and forest management become apparent, meeting subsistence needs alone is no longer adequate incentive for people's continued participation.

In recent years, the modes and design of decentralization in many countries have increasingly acknowledged the need to offer households, communities and local governments more attractive incentives and address broader livelihoods concerns. Local forest management and decentralization have thus been increasingly oriented towards securing rights to forest resources, sustaining livelihoods, and capturing and sharing revenues from commercial activities.

In some countries, state monopolies on forest product harvesting and marketing, and prohibitions on sale of products and transfers of rights are being relaxed or dismantled. Since the 1990s, China has allowed households to sell wood in the free market rather than to state monopolies, as during the previous 40 years. Further reforms underway would reduce the revenue share of the state and increase that of tree-growing households from its current share of 20 per cent of gross revenues. Several states in India are similarly increasing the benefit shares of village-based joint forest management groups. In Indonesia the sharing of revenues between central and local governments continues to be the subject of negotiation.

Households and forest user groups are also organizing themselves to take advantage of opportunities that open up. Some attempt to tap into niche markets through certification, green marketing, ecotourism and other channels. Partly in reaction to inequities and bureaucratic inefficiencies, many customary forest owners in Papua New Guinea are beginning to invest their own resources to develop needed capacities and procure inputs and technical assistance in a bid to enhance returns from their forest resources (see Box 3.1). In South Asia, local forest management groups are coalescing and organizing to achieve commercially viable scales of operation through alliances, cooperatives, federations and joint ventures for forest plantations and product processing (Capistrano, 2000; Bahuguna et al, 2004; Saigal, 2005; Boaz and Boaz, 2006).

An increasing number reinvest their forest revenues not only in forestry but also in local development and social welfare projects. In Nepal, after ploughing the required 25 per cent of revenues back into forest management, many forest user groups construct schools, irrigation canals and other local infrastructure. They also use a significant portion to help expand livelihood options for the poorest members of their communities (Shrestha, 2000; Chapter 5 of this volume). Joint forest management committees and forest cooperatives in India similarly allocate portions of collective funds for village development and social safety nets for poor households. In China, similar arrangements have begun to be mandated. In 2003, China's Jiangxi Province allowed collective forests to be leased to individuals and corporate entities. Of the fees earned, 70 per cent is to be distributed to members of the collective and the rest (30 per cent) can be used for village public welfare and development needs (Hyde et al, 2003; Chapter 6 of this volume).

These are promising trends. However, most countries will have to do more to make forest management under decentralization a more attractive and viable proposition. At the minimum, the package of incentives will have to include the provision of secure legal rights to forest lands and resources. In addition, the package must include at least some of the following: deregulation and streamlining of bureaucratic requirements to reduce transaction costs; simplified taxation; more equitable production and revenue- and benefit-sharing arrangements; provision of adequate infrastructure; access to credit, technical support and markets; and mechanisms to reduce risks.

Concluding Comments

There are indeed promising and hopeful signs that decentralized forest governance can work under certain conditions. Degraded forest lands are regenerating under various forms of individual household and group management. Capacities and institutions are being built and strengthened, and confidence and skills are being developed at different levels among actors and stakeholders. But it is difficult to gauge progress made under decentralization, and it is even more difficult to sustain decentralization initiatives. The goalposts for decentralization have been changing constantly as new, more ambitious objectives and expectations have been added and as the initial goals have been redefined over time.

Decentralization is expected to yield a combination of public and private goods, produce multiple products, provide environmental services and biodiversity conservation, encourage allocative efficiency, enhance social justice and equity, improve governance overall, and be economically viable as well. The evolving goals

Box 13.1 Major issues identified by thematic breakout groups (equity, legality and finance) during the workshop, 4–6 September 2006

- In most Asian countries, the involvement of women in the development process is largely ignored; decision-making and benefit sharing in forestry are controlled by local elites. Gender mainstreaming and capacity-building would improve women's participation in forestry.
- Government officials feel accountable to higher authorities but not to the communities. Performance evaluations should consider officials' contributions to community empowerment.
- Decentralization in the forestry sector is constrained by funding and capacity gaps, insufficient transfer of authority, and unclear roles and responsibilities at various levels.
- Lack of trust between government and communities and within the government bureaucratic layers accounts for the poor level of commitment to decentralization.
- Laws and regulations are highly inconsistent and frequently changing.
- A clear legal framework is necessary to ensure that agreed arrangements in decentralization are sound and enduring.
- In most cases in Asia, ethnic minority communities are marginalized and lack rights to their land. Equity in benefit sharing is not always addressed and is often left unclear.
- Logging bans are not observed in practice; rampant illegal logging is both a cause and an effect of corruption in most Asian countries. As it is so pervasive, the solution lies beyond intervention in the forestry sector.
- In many cases, customary institutions are good mechanisms for managing forests. Institutional restructuring could help improve implementation of decentralization.
- Decentralization presents both threats (decentralization of corruption) and opportunities (local community control and empowerment).
- Better governance in forestry requires greater transparency through the establishment of new laws and increased media scrutiny.

and expectations of decentralization reflect the general evolution in development and administration, and changing ideals for environmental resource governance. As many analysts have noted, decentralization is not a panacea. The empirical evidence also suggests that decentralization cannot guarantee better governance and better forest management. In many cases, forest conditions can also worsen (for example in Japan, Korea, Switzerland, New Zealand, India and Nepal) before they get better.

The many goals and expectations from decentralization can be contradictory and may require trade-offs. In reality, the space for the proverbial win–win outcomes tends to be narrower than initially supposed. Widening this space will entail significant transaction costs and require altogether different values and new ways of working and making decisions. It could take time, and there will still be winners and losers in the end. But people and institutions learn and adapt all the time. They constantly renegotiate and reshape their thinking and institutional arrangements to better balance or accommodate competing views based on lessons from the past. Herein lies the hope that although it could be complicated and take time, decentralization will lead to improvements that have not yet been achieved under more centralized modes of governance.

Notes

1 Public forest refers to those forests which are managed primarily for ecological benefit; hence they are strictly protected, and logging and collection of other forest products is heavily regulated.
2 Economic forests are managed mainly for the production of timber, fuelwood and other forest products.
3 Privatization in Vietnam includes forest management by individual households and state and joint venture enterprises, where the forest is allocated to the owner for long-term management (50 years).

References

Bahuguna, V. K., Mitra, K., Capistrano, D. and Saigal, S. (eds) (2004) *Root to Canopy: Regenerating Forests through Community–State Partnerships*, Winrock International – India and Commonwealth Forestry Association, New Delhi

Banerjee, A. K. (2004) 'Tracing social initiatives towards JFM', in V. K. Bahuguna, K. Mitra, D. Capistrano and S. Saigal (eds) *Root to Canopy: Regenerating Forests through Community–State Partnerships*, Winrock International – India and Commonwealth Forestry Association, New Delhi

Barber, C. V. and Scheweithelm, J. (2000) *Trial by Fire: Forest Fires and Forestry Policy in Indonesia's Era of Crisis and Reform*, World Resources Institute, Washington, DC

Barr, C., Resosudarmo, I. A. P., Dermawan, A., McCarthy, J. F., Moeliono, M. and Setiono, B. (eds) (2006) *Decentralization of Forest Administration in Indonesia: Implications for Forest Sustainability, Economic Development and Community Livelihoods,* Center for International Forestry Research, Bogor, Indonesia

Boaz, O. and Boaz, A. (2006) 'The decentralized approach to forest management – A case study from Central India', paper presented at the regional workshop on Forest Governance and Decentralization in Asia and the Pacific, 4–6 September 2006. Yogyakarta, Indonesia

Brown, C. and Durst, P. (2003) *State of Forestry in Asia and the Pacific – 2003,* Asia-Pacific Forestry Commission, Food and Agriculture Organization of the United Nations, Bangkok

Capistrano, A. D. (1994) 'Tropical forest depletion and the changing macroeconomy', in K. Brown and D. Pearce (eds) *The Causes of Tropical Deforestation*, UCL Press, London

Capistrano, D. (2000) 'Non-timber forest product-based enterprise in forest conservation and community development: India's evolving institutional context', in B. Krishnapillay et al (eds) *Forests and Society: The Role of Research*, XXI International Union of Forest Research Organizations (IUFRO) World Congress, Sub-plenary Session 1, pp431–440

Capistrano, D. and Colfer, C. J. P. (2005) 'Decentralization: Issues, lessons and reflections', in C. J. P. Colfer and D. Capistrano (eds) *The Politics of Decentralization: Forests, Power and People*, Earthscan, London

Carter, J. and Gronow, J. (2005): *Recent Experience in Collaborative Forest Management. A Review Paper*, CIFOR Occasional Paper No 43, Center for International Forestry Research, Bogor, Indonesia

Cleaver, F. (2005) 'The inequality of social capital and the reproduction of chronic poverty', *World Development,* vol 33, no 6, pp893–906

Colchester, M. with Boscolo, M., Contreras-Hermosilla, A., Del Gatto, F., Dempsey, J., Lescuyer, G., Obidzinski, K., Pommier, D., Richards, M., Sembiring, S. N., Tacconi, L., Vargas Rios, M. T. and Wells, A. (2006) *Justice in the Forest: Rural Livelihoods and Forest Law Enforcement,* Center for International Forestry Research, Bogor, Indonesia

Colfer, C. J. P. and Capistrano, D. (eds) (2005) *The Politics of Decentralization: Forests, Power and People*, Earthscan, London

Colfer, C. J. P. (2005) *The Complex Forest: Communities, Uncertainty and Adaptive Collaborative Management,* Resources for the Future, Washington, DC, and Center for International Forestry Research, Bogor, Indonesia

Contreras-Hermosilla, A. and Fay, C. (2005) *Strengthening Forest Management in Indonesia through Land Tenure Reform: Issues and Framework for Action,* Forest Trends, Washington, DC, and Basic Books, New York

Durst, P. B., Brown, C., Tacio, H. D. and Ishikawa, M. (2005) *In Search of Excellence: Exemplary Forest Management in Asia and the Pacific,* FAO and Regional Community Forestry Training Center, Bangkok

Enters, T., Victor, M. and Durst, P. B. (eds) (2000) *Decentralization and Devolution of Forest Management in Asia and the Pacific,* RECOFTC Report No 18 and RAP Publication 2000/1, Regional Community Forestry Training Center, Bangkok

Edmunds, D. and Wollenberg, E. (2003) *Local Forest Management: The Impacts of Devolution Policies*, Earthscan, London

Ferguson, I. and Chandrasekharan, C. (2005) 'Paths and pitfalls of decentralization for sustainable forest management: Experiences of the Asia-Pacific region', in C. J. P. Colfer and D. Capistrano (eds) *The Politics of Decentralization: Forests, Power and People*, Earthscan, London

FAO/ITTO (Food and Agriculture Organization of the United Nations and International Tropical Timber Organization) (2005) *Best Practices for Improving Law Compliance in the Forestry Sector*, FAO and ITTO, Rome

FAO (2006) *Global Forest Resources Assessment 2005 – Progress Towards Sustainable Forest Management*, FAO Publications, Rome

Gibson, C. C., McKean, M. A. and Ostrom, E. (2000) *People and Forests: Communities, Institutions and Governance*, MIT Press, Cambridge, MA

Glück, P., Rayner, J., Cashore, B., Agrawal, A., Bernstein, S., Capistrano, D., Hogl, K., Liss, B. M., McDermott, C., Maini, J. S., Oksanen, T., Ollonqvist, P., Pülzl, H., Rametsteiner, E. and Pleschberger, W. (2005) 'Changes in the governance of forest resources', in G. Mery, R. Alfaro, M. Kanninen and M. Lobovikov (eds) *Forests in the Global Balance – Changing Paradigms*, IUFRO World Series Volume 17, International Union of Forest Research Organizations, Helsinki

Hyde, W. F., Belcher, B. and Xu, J. (eds) (2003) *China's Forests: Global Lessons from Market Reforms*, Resources for the Future, Washington, DC, and Center for International Policy Research, Bogor, Indonesia

Küchli, C. and Blaser, J. (2005) 'Forests and decentralization in Switzerland: A sampling', in C. J. P. Colfer and D. Capistrano (eds) *The Politics of Decentralization: Forests, Power and People*, Earthscan, London

Lao PDR (2005) *Forestry Strategy to the Year 2020 of Lao PDR*, Vientiane, Lao PDR

Larson, A. M. and Ribot, J. C. (2005) 'Democratic decentralization through a natural resource lens: An introduction', in J. C. Ribot and A. M. Larson (eds) *Democratic Decentralization through a Natural Resource Lens: Cases from Africa, Asia and Latin America*, Routledge, London

Lynch, O. J. (2005) 'Concepts and strategies for promoting legal recognition of community-based property rights: Insights from the Philippines and other nations', in P. Brosius, A. L. Tsing and C. Zerner (eds) *Communities and Conservation: Histories and Politics of Community-Based Natural Resource Management*, AltaMira Press, Walnut Creek, CA

Lynch, O. J. and Harwell, E. (2002) *Whose Natural Resources? Whose Common Good?*, Center for International Environmental Law, Washington, DC

Mayers, J. and Bass, S. (1999) *Policy that Works for Forests and People*, International Institute for Environment and Development, London

Molnar, A., Scherr, S. and Khare, A. (2004) *Who Conserves the World's Forests? Community-driven Strategies to Protect Forests and Respect Rights*, Forest Trends, Washington, DC

Nguyan, T. Q. (2005) *What Benefits and for Whom? Effects of Devolution of Forest Management in Dak Lak, Vietnam*, Shaker Verlag, Berlin, Germany

Obidzinski, K., Andrianto, A., Hernawan, D., Belle, C. W. and Kusumawati, R. (2006) *Illegal Forest Activities in Berau and Kutai Timur: Impacts, Driving Forces and Remedies*, Governance Brief No 26, Center for International Forestry Research, Bogor, Indonesia

Ostrom, E., Dietz, T., Dolsak, N., Stern, P. C., Stonich, S. and Weber, E. U. (eds) (2002) *The Drama of the Commons*, National Academy Press, Washington, DC
Peluso, N. L. and Watts M. (eds) (2001) *Violent Environments*, Cornell University Press, Ithaca, NY, and London
Rangachari, C. S. and Mukherji, S. D. (2000) *Contemporary Issues in Indian Forestry: Old Roots New Shoots*, Winrock International and Ford Foundation, New Delhi
Ravindranath, N. H., Sudha, P. Malhotra, K. C. and Palit, S. (2004) 'Sustaining JFM in India', in N. H. Ravindranath and P. Sudha (eds) *Joint Forest Management in India: Spread, Performance and Impact*, Universities Press, Bangalore, India
Resosudarmo, I. A. P. and Dermawan, A. (2002) 'Forests and regional autonomy: The challenge of sharing the profits and pains" in C. J. P. Colfer and I. A. P. Resosudarmo (eds) *Which Way Forward? People, Forests and Policymaking in Indonesia*, Resources for the Future/Center for International Forestry Research, Washington, DC
Ribot, J. C. (2005) 'Choosing representation: Institutions and powers for decentralized natural resource management', in C. J. P. Colfer and D. Capistrano (eds) *The Politics of Decentralization: Forests, Power and People*, Earthscan, London
Ritchie, B. and Haggith, M. (2005) 'The push-me, pull-you of forest devolution in Scotland', in C. J. P. Colfer and D. Capistrano (eds) *The Politics of Decentralization: Forests, Power and People*, Earthscan, London
Saigal, S. (2003) 'Joint management of state forest lands: Experiences from India', in M. Garforth and J. Mayers (eds) *Plantations, Privatization, Poverty and Power*, Earthscan, London
Saigal, S., Mitra, K. and Lal, P. (2005) 'Empowering the forest-dependent poor in India', in J. A. Sayer and S. Maginnis (eds) *Forests in Landscapes: Ecosystem Approaches to Sustainability*, Earthscan/IUCN, London
Saway, A. and Mirasol, F. (2005) 'Decentralizing protected area management at Mount Kitanglad', in C. J. P. Colfer and D. Capistrano (eds) *The Politics of Decentralization: Forests, Power and People*, Earthscan, London
Sayer, J., Elliott, C., Barrow, E., Gretzinger, S., Maginnis, S., McShane, T. and Shepherd, G. (2005) 'Implications for biodiversity conservation of decentralized forest resource management', in C. J. P. Colfer and D. Capistrano (eds) *The Politics of Decentralization: Forests, Power and People*, Earthscan, London
Sayer, J. A. and Maginnis, S. (eds) (2005) *Forests in Landscapes: Ecosystem Approaches to Sustainability*, Earthscan/IUCN, London
Scherr, S., White, A. and Kaimowitz, D. (2004) *A New Agenda for Forest Conservation and Poverty Reduction: Making Markets Work for Low-income Producers*, Forest Trends, Washington, DC
Setiono, B. and Husein, Y. (2005) *Fighting Forest Crime and Promoting Prudent Banking for Sustainable Forest Management – The Anti Money Laundering Approach*, CIFOR Occasional Paper No 44, Center for International Policy Research, Bogor, Indonesia
Shrestha, B. K. (2000) *Good Governance in Nepal: Perspectives from Panchathar and Chanchanpur Districts*, Rural Development Foundation, Kathmandu
Singh, T. P. and Varalakshmi, V. (eds) (1998) *The Decade and Beyond: Evolving Community-State Partnership*, Tata Energy Research Institute, New Delhi
Siswanto, W. and Wardojo, W. (2005) 'Decentralization of the forestry sector: Indonesia's experience', in C. J. P. Colfer and D. Capistrano (eds) *The Politics of Decentralization: Forests, Power and People*, Earthscan, London

Tacconi, L., Boscolo, M. and Brack, D. (2003) *National and International Policies to Control Illegal Forest Activities*, Center for International Forestry Research, Bogor, Indonesia

White, A. and Martin, A. (2002) *Who Owns the World's Forests? Forest Tenure and Public Forests and Transition*, Forest Trends and Center for International Environmental Law, Washington, DC

Xu, J. and Hyde, W. (2005) 'From centrally planned economy to vigorous rural enterprise: China', in M. Garforth and J. Mayers (eds) *Plantations, Privatization, Poverty and Power*, Earthscan, London

Yin, R. (2003) 'Central characteristics of reform: Measures of the effects of improved property rights, a stable policy environment and environmental protection', in W. F. Hyde, B. Belcher and J. Xu (eds) *China's Forests: Global Lessons from Market Reforms*, Resources for the Future, Washington, DC, and Center for International Policy Research, Bogor, Indonesia

Afterword

R. J. Fisher

Decentralization and devolution have been a major policy theme in forest governance in Asia and the Pacific for well over a decade. This has been reflected in a series of conferences that began with a seminar in Davao in 1998 (published as Enters et al, 2000). A second major conference, held in Interlaken in 2004 (Colfer and Capistrano, 2005), looked beyond Asia and the Pacific but presented a number of cases from the region. This book reflects the findings of the third conference, held in Yogyakarta in September 2006, which again focused on Asia and the Pacific.

I do not propose to review or summarize this rich collection, which analyses experiences from various countries in the region. This has already been done very effectively in the concluding chapter. Instead, I would like to briefly explore what has changed, and what has not changed, since the Davao seminar. I will focus on devolution to the community level, although this book goes beyond devolution to communities and looks at the processes of decentralization as they affect various levels of government – national, provincial and district.

The Davao seminar presented a wide-ranging review of the progress of decentralization and devolution of forest management in the region. Although I found the discussion and debate stimulating, as a participant I was left with a strong sense of dissatisfaction about the limited extent to which meaningful devolution of the authority and power over forest management had occurred in practice. It is obvious from the research presented in this volume that not as much has changed in the intervening eight years as optimists and advocates of devolution might have hoped, and that real shifts of power remain limited nearly a decade after Davao. Guiang, Esguerra and Bacalla (Chapter 11) note a 'growing "disconnect" between what was intended and what is actually happening in forest management in the Philippines'. One of the contributions of this book is that it documents the nature of this disconnect.

The introduction to the Davao proceedings (Fisher et al, 2000) noted that 'the tendency [in the Asia-Pacific region] has been to grant local communities the responsibility for protecting forest resources, without granting them the rights to use them in a major way'. Related to this was 'the decentralization of responsibility without devolution of the power to make independent decisions or to take action outside narrow parameters set by forest authorities'. In other words, the objectives of forest management are set by outsiders and communities are generally not permitted to decide how to manage forests or what to manage them for. This provides little incentive to take an active role in management or protection.

Those points could easily have been incorporated in a summary of the current volume. So why has there been so little change?

There are several reasons why governments and forest authorities have been reluctant to devolve decision-making authority over forest management to communities. One of these is a genuinely felt lack of trust in the capacity of communities to manage forests sustainably. Another, which is difficult to talk about except in general terms, is the existence of vested interests within forest departments, tied up with the desire to maintain the forestry profession's self-image, with concern that too much decentralization will reduce professional opportunities and, sometimes, with opportunities for corruption. Another is a concern on the part of forest authorities that they are responsible for seeing that forests are managed in the national interest.

One of the results of the lack of trust and the concern for the national interest is that forest authorities define management practices in considerable detail, leaving communities to make only the most trivial decisions. The question that arises from this is whether it is necessary to have detailed guidelines to ensure sustainable management or the national interest. Are there other ways to maintain accountability?

Jesse Ribot, whose work on decentralization of management of natural resources is mentioned several times in this book, has argued for the application of a 'minimum standards approach' (Ribot, 2002). The idea is that effective management can be achieved by setting broad guidelines and short lists of what cannot be done rather than long lists of what must be done and how. People responsible for decentralized management are held accountable for the maintenance of these minimum standards. In terms of forest management, an example might be guidelines that require community managers to maintain a certain percentage of forest cover (or number of stems per hectare), perhaps avoiding the harvesting of specific species. Other matters, such as timing of harvesting, selection of preferred species and distribution of income from forest activities, can be left to communities to decide.

In Chapter 6, Liu Dachang, in reference to China, points out that ownership is not enough – power is also needed for meaningful decentralization. This is a very important point. It is useful to define power as the capacity to make effective decisions – that is, decisions that can be implemented in practice. The question is not whether governments or communities entirely control forests, but whether the domain within which communities can make decisions is significant, clear and enforceable. Power sharing is possible if the parameters are agreed and clear.

Frequently, the debate about who should control forests is framed in terms of who 'owns' or should own them. On one side it is argued that governments should own forests so that they can protect national or even global interests. Others argue that communities should own forests because they will invest time and effort in sustainable forest management if they have secure access to benefits, or sometimes because communities living in and near forests have a moral right to own them.

The problem with both those arguments is that the whole idea of ownership needs to be unpacked. In practice, ownership by the state does not necessarily preclude communities' or individuals' having clear and agreed rights to use products or to act in forests. Similarly, community ownership of forest areas or forest resources does not mean that they can do whatever they want with them. 'Ownership' is always constrained in various ways. Many writers on tenure have described tenure in terms of 'bundles of rights', a concept that implies combinations of rights and responsibilities. The issue, then, is not who owns forests, but which groups have effective power over decisions about how forests are used and how this power is shared.

It is easy to be pessimistic about the implications of this book, in terms of the limited progress towards decentralized management of forest that it documents. However, consider the increasing acknowledgement of the issues and the continuing shift away from uncritical advocacy of programmatic approaches and towards critical analysis of the complexities and realities of decentralized forest governance. It is often said that a pessimist sees that a glass is half empty and an optimist sees that it is half full. I think this book has a lot to offer the optimists.

References

Colfer, C. J. P. and Capistrano, D. (eds) (2005) *The Politics of Decentralization: Forests, Power and People*, Earthscan, London

Enters, T., Victor, M. and Durst, P. B. (eds) (2000) *Decentralization and Devolution of Forest Management in Asia and the Pacific*, RECOFTC Report 18 and RAP Publication 2000/1, Regional Community Forestry Training Center and FAO Regional Office for Asia and the Pacific, Bangkok

Fisher, R. J., Durst, P. B., Enters, T. and Victor, M. (2000) 'Overview of themes and issues in devolution and decentralization of forest management in Asia and the Pacific', in T. Enters, P. B. Durst and M. Victor (eds) *Decentralization and Devolution of Forest Management in Asia and the Pacific*, RECOFTC Report no 18 and RAP Publication 2000/1, RECOFTC and FAO Regional Office for Asia and the Pacific, Bangkok, Thailand, pp1–5

Ribot, J. (2002) *Democratic Decentralization of Natural Resources: Institutionalizing Popular Participation*, World Resources Institute, Washington, DC

Index

For Product Safety Concerns and Information please contact our EU
representative GPSR@taylorandfrancis.com
Taylor & Francis Verlag GmbH, Kaufingerstraße 24, 80331 München, Germany

www.ingramcontent.com/pod-product-compliance
Ingram Content Group UK Ltd.
Pitfield, Milton Keynes, MK11 3LW, UK
UKHW020012200926
475702UK00014B/1847

* 9 7 8 0 4 1 5 8 4 5 6 1 8 *